Quantum Dots - Theory and Applications

Quantum Dots - Theory and Applications

Editor

Rodolph Donovan

scitus academics

Quantum Dots - Theory and Applications
Edited by **Rodolph Donovan**

ISBN: 978-1-68117-320-7
Library of Congress Control Number: 2016937631

SCITUS Academics LLC,
www.scitusacademics.com
Box No. 4766, 616 Corporate Way,
Suite 2, Valley Cottage,
NY 10989

Notice

Printed in United States of America on Acid Free Paper

Preface

A quantum dot is a nanoparticle made of any semiconductor material such as silicone, cadmium selenide, cadmium sulfide or arsenide. Quantum dots can increase the energy efficiency of solar cells. A light photon generates one electron in normal solar cells. Experiments with both silicone quantum dots and lead sulfide quantum dots can produce two electrons for one light photon. The use of quantity points in solar cells could therefore significantly increase their efficiency in the production of electricity. Quantum dots (QD) are semiconductor nanoscale devices that contain electrons or electron holes in all three spatial dimensions. They can be done via various possible routes, including colloidal synthesis, plasma synthesis or mechanical manufacturing. The electronic properties of quantum dots fall between those of bulk semiconductors and those of discrete molecules of comparable size and optoelectronic properties such as band gap can be tuned to a particular composition according to the particle size and shape. Due to the high tunability of the properties, QDs are important in many research applications such as transistors, solar cells, LEDs and lasers for diode. For example, QDs' ability to convert and adjust a spectrum accurately makes it ideal for LCD display. Previous LCD displays can waste energy into a more balanced light by converting poor red-green, rich blue-yellow white light. Only the necessary colors for ideal images are contained in the screen by using QDs. The result is a lighter, clearer and more energy efficient screen. Researchers also work on the use of quantum dots in displays for applications ranging from mobile phones to large-screen TVs that consume less power than current displays. The red, green and blue colors used to generate the full spectrum of colors would be available by placing different quantum dots in each pixel of the display screen.

The book covers some new research and innovative developments in the field of fundamental excitations, photoluminescence experiments, epitaxial techniques, decoherence, charging states in the field of quantum dot devices.

Table of Contents

Chapter 1	Theory of Excitons and Excitonic Quasimolecules Formed from Spatially Separated Electrons and Holes in Quasi-Zero-Dimensional Nanostructures 1
Chapter 2	Water-Soluble Silicon Quantum Dots with Quasi-Blue Emission 35
Chapter 3	Nitrogen-Doped Carbon Dots for "green" Quantum Dot Solar Cells 47
Chapter 4	Theoretical and experimental studies of (In,Ga)As/GaP quantum dots 63
Chapter 5	Quantum Dots Prepared by Droplet Epitaxial Method 77
Chapter 6	The Preparation and Application of Dendrimer Modified CdTe/CdS Near Infrared Quantum Dots for Brain Cancer Cells Imaging 123
Chapter 7	Semiconductor Quantum Dots for Biomedicial Applications 135

Chapter 8 **Physical Reasons of Emission Varying in CdSe/ZnS and CdSeTe/ZnS Quantum Dots at Bioconjugation to Antibodies** **161**

Chapter 9 **Growth of InAs Quantum Dots on Germanium Substrate Using Metal Organic Chemical Vapor Deposition Technique** **203**

Chapter 10 **Quantum Dots as Multifunctional Materials for Tumor Imaging and Therapy** **221**

Chapter 11 **Fabrication of low-density GaN/AlN quantum dots via GaN thermal decomposition in MOCVD** **247**

Chapter 12 **Assembly of CdS Quantum Dots onto Hierarchical TiO2 Structure for Quantum Dots Sensitized Solar Cell Applications** **259**

Chapter 13 **Investigation of the InAs/GaAs Quantum Dots' Size: Dependence on the Strain Reducing Layer's Position** **277**

Index **295**

1

CHAPTER

THEORY OF EXCITONS AND EXCITONIC QUASIMOLECULES FORMED FROM SPATIALLY SEPARATED ELECTRONS AND HOLES IN QUASI-ZERO-DIMENSIONAL NANOSTRUCTURES

Sergey I. Pokutnyi[1] and Włodzimierz Salejda[1]

[1] *Chuiko Institute of Surface Chemistry, National Academy of Sciences of Ukraine, Kyiv, Ukraine*

[2] *Technical University of Wroclaw, Wroclaw, Poland*

1. INTRODUCTION

Quasi-zero-dimensional semiconductor nanosystems consisting of spherical semiconductor nanocrystals, i.e., quantum dots with radii of a =1-10 nm and containing cadmium sulphide and selenide, gallium arsenide, germanium, silicon, and zinc selenide in their volume, and synthesized in a borosilicate glass matrix currently attract particular research attention due to their unique photoluminescent properties, i.e., the ability to efficiently emit light in the visible or near infrared ranges at room temperature [1-10]. The optical and electro-optical properties of such quasi-zero dimensional nanosystems are to a large extent controlled by the energy spectrum of the spatially confined electron-hole pair (exciton) [4-16].

In most theoretical models for calculating the energy spectra of quasiparticles in quantum dots (QDs), the effective mass approximation is used, which was considered applicable for QDs by analogy with bulk single crystals [11-13]. However, the problem concerning the applicability of the effective mass approximation to the description of semiconductor QDs remains unsolved [4-18].

In [14], a new modified effective mass method is proposed to describe the exciton energy spectrum in semiconductor QDs with radii of $a \approx a_{ex}$ (a_{ex} is the exciton Bohr radius in the semiconductor material contained in the QD volume). It was shown that, within a model in which the QD is represented as an infinitely deep potential well, the effective mass approximation can be applied to the description of an exciton in QDs with radii a comparable to the exciton Bohr radius a_{ex}, assuming that the reduced effective exciton mass is a function of the radius a, $\mu = \mu(a)$.

In the adiabatic approximation and within the modified effective mass method [14], an expression for the binding energy of an exciton, whose electron and hole move within the semiconductor QD volume, was derived in [15]. In [15], the effect of significantly increasing the exciton binding energy in cadmium selenide and sulphide QDs with radii a, comparable to the exciton Bohr radii a_{ex} and relative to the exciton- binding energy in cadmium selenide and sulphide single crystals (by factors of 7.4 and 4.5, respectively) was also detected.

In the experimental study [7], it was found that excess electrons produced during interband excitation of the cadmium sulphide QD had a finite probability of overcoming the potential barrier and penetrating into the borosilicate glass matrix, into which the QD is immersed. In experimental studies [10, 19] (as well as in [7]) using glass samples with cadmium-sulphide and zinc selenide QDs, it was found that the electron can be localized in the polarization well near the outer QD surface, while the hole moves within the QD volume.

In [10, 19], the optical properties of borosilicate glass samples containing QD zinc selenide were experimentally studied. The average radii of such QDs were in the range $a \approx 2.0$-4.8 nm. In this case, the values of a are comparable to the exciton Bohr radius $a_{ex} \approx$ 3.7 nm in a ZnSe single crystal. At low QD concentrations, when the optical properties of the samples are mainly controlled by those of individual QDs in the borosilicate glass matrix, a shift of the peak of the low temperature luminescence spectrum to the short wavelength region (with respect to the band gap E_g of the zinc selenide-single crystal) was observed. The authors of [10] assumed that this shift was caused by quantum confinement of the energy spectra of the electron and the hole localized near the spherical surface of the QD. In this case, the following problem remained open: the quantum confinement of the state of which electron and hole (the hole moving in the QD volume and the electron localized at the outer spherical QD-dielectric matrix interface or the electron and hole localized in the QD volume) caused such a shift in the luminescence spectrum peak?

The use of semiconductor nanosystems as the active region of nanolasers is prevented by the low binding energy of the QD exciton [8, 9, 13]. Therefore, studies directed at the search for nanostructures in which a significant increase in the binding energy of QD excitons can be observed are of importance.

Currently, the theory of exciton states in quasi- zero- dimensional semiconductor nanosystems has not been adequately studied. In particular, no theory exists for an exciton with a spatially separated electron and hole in quasi- zero- dimensional nanosystems. Therefore, in this study, we developed the theory of an exciton formed from a spatially separated electron and hole (the hole is in the semiconductor QD volume and the electron is localized at the outer spherical surface of the QD-dielectric matrix interface) [20-22]. It was shown that the short wavelength shift of the peak of the low temperature luminescence spectrum of samples containing zinc selenide QDs, observed under the experimental conditions of [10], was caused by quantum confinement of the ground state energy of the exciton with a spatially separated electron and hole. The effect of significantly increasing the binding energy of an exciton (with a spatially separated electron and hole) in a nanosystem containing zinc selenide QDs, compared with the binding energy of an exciton in a zinc selenide single crystal (by a factor of 4.1-72.6), was detected [20-22].

In [10, 19], a shift of the spectral peak of the low temperature luminescence was also observed for samples with a QD concentrations from $x = 0.003$-1%. It was noted [10, 19] that at such a QD content in the samples, the interaction between charge carriers localized above the QD surfaces must be taken into account. Therefore, in [23, 24], we develop the theory of excitonic quasimolecules (biexcitons) (formed from spatially separated electrons and holes) in a nanosystem, which consists of ZnSe QDs synthesized in a borosilicate glass matrix.

2. SPECTROSCOPY OF EXCITONS IN QUASI - ZERO - DIMENSIONAL NANOSYSTEMS

Let us consider the simple model of a quasi-zero-dimensional system, i.e., a neutral spherical semiconductor QD of the radius a, which contains semiconductor material with the permittivity ε2 in its volume, surrounded by a dielectric matrix with permittivity ε_1. A hole h with the effective mass m_h moves in the QD volume, while an electron e with the effective mass $m_e^{(1)}$ lies in the matrix (r_e and r_h are the distances from the QD centre to the electron and hole). Let us assume that the QD valence band is parabolic. Let us also assume that there is an infinitely high potential barrier at the spherical QD-dielectric matrix interface; therefore, the hole h cannot leave the QD volume and the electron e cannot penetrate into the QD volume in the model under study [20-22].

The characteristic dimensions of the problem are the quantities:

$$a_h = \varepsilon_2 \hbar^2 / m_h e^2, \quad a_{ex} = \varepsilon_2 \hbar^2 / \mu e^2, \quad a_e = \varepsilon_1 \hbar^2 / m_e^{(1)} e^2, \tag{1}$$

where a_h and a_{ex} are the hole and exciton Bohr radii in the semiconductor with the permittivity ε2, e is the elementary charge, $\mu = m_e^{(2)} m_h / (m_e^{(2)} + m_h)$ is the reduced effective mass of the exciton, $m_e^{(2)}$ is the effective mass of an electron in the semiconductor with permittivity ε_2 and a_e is the electron Bohr radius in the dielectric matrix with the permittivity ε1. The fact that all characteristic dimensions of the problem are significantly larger than the interatomic distances a_0,

$$a, a_e, a_h, a_{ex} >> a_0$$

allows us to consider the electron and hole motion in the quasi-zero-dimensional nanosystem in the effective mass approximation [11-13].

We analysed the conditions of carrier localization in the vicinity of a spherical dielectric particle of the radius *a* with the permittivity ε_2 in [25-27]. In this instance, the problem of the field induced by the carrier near a dielectric particle immersed in a dielectric medium with the permittivity ε_1 was solved in a final analytical form and analytical expressions for the potential energy of the interaction of the carrier with the spherical interface of two media are presented.

Solving the Poisson equation with usual electrostatic boundary conditions

$$\phi(r', j \mid \mathbf{r}, i)\big|_{r'=a} = \phi(\mathbf{r}', j)\big|_{r'=a},$$
$$\varepsilon_1 \frac{\partial \phi(\mathbf{r}', j \mid \mathbf{r}, i)}{\partial \mathbf{r}'}\bigg|_{r'=a} = \varepsilon_2 \frac{\partial \phi(\mathbf{r}', j)}{\partial \mathbf{r}'}\bigg|_{r'=a}, \quad (2)$$

the potential $\varphi(\mathbf{r}', j|\mathbf{r}, i)|$ φ(r',j|r,i)| at the observation point $\mathbf{r}'$ in a medium with the permittivity ε_j, induced by the charge *e* at the point $\mathbf{r}$ in a medium with the permittivity ε_i, can be presented as a sum of the potentials induced by the image point charge $e'(\mathbf{r}_{ij}\,|\mathbf{r})$ at the point $\mathbf{r}_{ij} = (a/r)^2\mathbf{r}\delta_{ij} + \mathbf{r}(1 - \delta ij)$ and the linear distribution with the density $\rho_{ij}(y, r)$ of the image charge along a straight line passing through the centre of the dielectric particle with the radius a and the charge at the point $\boldsymbol{r}$ [25-27]:

$$\phi(\boldsymbol{r}',j\,|\,\boldsymbol{r},i)=\frac{e}{\varepsilon_j\,|\boldsymbol{r}'-\boldsymbol{r}|}+\frac{e'(\boldsymbol{r}_{ij}\,|\,\boldsymbol{r})}{\varepsilon_j\,|\boldsymbol{r}'-\boldsymbol{r}_{ij}|}+\frac{1}{\varepsilon_j}\int_0^{\infty}\frac{dy\rho_{ij}(y,\boldsymbol{r})}{|\boldsymbol{r}'-y(\boldsymbol{r}/r)|}$$

where

$$\mathrm{r}_{11}=(a^2/r^2)\boldsymbol{r},e'(\mathrm{r}_{11}\,|\,\mathrm{r})=-\beta(a/r)e \qquad \text{a}$$

$$\rho_{11}(y,r)=\beta\alpha\left(a^2/ry\right)^{(1-\alpha)}(e/a)\theta((a^2/r)-y);$$

$$\mathrm{r}_{22}=\left(a^2/r^2\right)\mathrm{r},e'\left(\mathrm{r}_{22}\,|\,\mathrm{r}\right)=\beta\left(a/r\right)e \qquad \text{b}$$

$$\rho_{22}(y,r)=\beta\left(1-\alpha\right)\left(a^2/ry\right)^{\alpha}\left(e/a\right)\theta\left(y-\left(a^2/r\right)\right);$$

$$r_{12}=r,e'(\mathrm{r}_{12}\,|\,\mathrm{r})=\beta e \qquad \text{c}$$

$$\rho_{12}(y,r)=\beta(1-\alpha)(r/y)^{\alpha}(e/r)\theta(y-r);$$

$$r_{21}=r,e'\left(r_{12}\,|\,r\right)=-\beta e \qquad \text{d}$$

$$\rho_{21}(y,r)=\beta\alpha\left(r/y\right)^{(1-\alpha)}(e/r)\theta\left(r-y\right) \qquad (3)$$

where $\theta(x)$ is the Heaviside unit-step function,

$$\beta=\frac{\left(\varepsilon_2-\varepsilon_1\right)}{\left(\varepsilon_2+\varepsilon_1\right)},\alpha=\frac{\varepsilon_1}{\left(\varepsilon_2+\varepsilon_1\right)} \qquad (4)$$

Using expressions (3)-(3d), the energy U(r e ,r h ,a) U(re,rh,a) of the polarization interaction of the electron and hole with the spherical QD-matrix interface at the relative permittivity $\varepsilon = (\varepsilon_2/\varepsilon_1) \gg\gg 1$ can be presented as an algebraic sum of the energies of the interaction of the hole and electron with self- $V hh'$ (r h ,a) Vhh'(rh,a) , $V_{ee'}(r_e, a)$ and "foreign" $V_{eh'}(\boldsymbol{r}_e, \boldsymbol{r}_h, a)$, $V_{he'}(\boldsymbol{r}_e, \boldsymbol{r}_h, a)$ images, respectively [15, 16, 26-28]:

$$U(\mathrm{r}_e,\mathrm{r}_h,a,\varepsilon)=V_{hh'}\left(r_h,a,\varepsilon\right)+V_{ee'}\left(r_e,a,\varepsilon\right)+V_{eh'}\left(\mathrm{r}_e,\mathrm{r}_h,a,\varepsilon\right)+V_{he'}\left(\mathrm{r}_e,\mathrm{r}_h,a,\varepsilon\right), \qquad (5)$$

Where

$$V_{hh'}\left(r_h,a,\varepsilon\right)=\frac{e^2\beta}{2\varepsilon_2 a}\left(\frac{a^2}{a^2-r_h^2}+\varepsilon\right), \tag{6}$$

$$V_{ee'}\left(r_e,a,\varepsilon\right)=-\frac{e^2\beta}{2\varepsilon_1 a}\cdot\frac{a^4}{r_e^2(r_e^2-a^2)}, \tag{7}$$

$$V_{he'}\left(\mathrm{r_e},\mathrm{r_h},a,\varepsilon\right)=\frac{e^2\beta}{2\varepsilon_2 a}\cdot\frac{a^2}{r_e\left|\mathrm{r_h}-\left(a/r_e\right)^2\mathrm{r_e}\right|}, \tag{8}$$

$$V_{eh'}\left(\mathrm{r_e},\mathrm{r_h},a,\varepsilon\right)=-\frac{e^2\beta}{2\varepsilon_1 a}\cdot\frac{a^2}{r_h\left|\mathrm{r_e}-\left(a/r_h\right)^2\mathrm{r_h}\right|}. \tag{9}$$

In the studied simple model of a quasi-zero-dimensional nanostructure within the above approximations and in the effective mass approximation using the triangular coordinate system [14-16], $r_e = |\boldsymbol{r}_e|$, $r_h = |\boldsymbol{r}_h|$, $r = |\boldsymbol{r}_e - \boldsymbol{r}_h|$, with the origin at the centre of the QD, the exciton Hamiltonian (with a spatially separated hole moving within the QD volume and an electron in the dielectric matrix) takes the following form [20-22, 29-32]:

$$\begin{aligned}H\left(\mathrm{r_e},\mathrm{r_h},\mathrm{r},a\right)=&-\frac{\hbar^2}{2m_e^{(1)}}\left(\frac{\partial^2}{\partial r_e^2}+\frac{2}{r_e}\cdot\frac{\partial}{\partial r_e}+\frac{r_e^2-r_h^2+r^2}{r_e r}\cdot\frac{\partial^2}{\partial r_e\partial r}\right)-\\&-\frac{\hbar^2}{2m_h}\left(\frac{\partial^2}{\partial r_h^2}+\frac{2}{r_h}\cdot\frac{\partial}{\partial r_h}+\frac{r_h^2-r_e^2+r^2}{r_h r}\cdot\frac{\partial^2}{\partial r_h\partial r}\right)-\frac{\hbar^2}{2\mu_0}\left(\frac{\partial^2}{\partial r^2}+\frac{2}{r}\cdot\frac{\partial}{\partial r}\right)+\\&+V_{eh}\left(\mathrm{r}\right)+U\left(\mathrm{r_e},\mathrm{r_h},a,\varepsilon\right)+V_e\left(r_e\right)+V_h\left(r_h\right)+E_g,\end{aligned} \tag{10}$$

where the first three terms are the operators of the electron, hole and exciton kinetic energy, *Eg* is the band gap in the semiconductor with the permittivity ε_2 and μ_0 =

m (1) *e m h* /(*m* (1) *e* +*m h*) me(1)mh/(me(1)+mh) is the reduced effective mass of the exciton (with a spatially separated hole and electron). In the exciton Hamiltonian (10), the polarization interaction energy $U(\mathbf{r}_e, \mathbf{r}_h, a, \varepsilon)$ (5) is defined by formulas (6)-(9) and the electron-hole Coulomb interaction energy *Veh*(*r*) is described by the following formula:

$$V_{eh}(r) = -\frac{1}{2}\left(\frac{1}{\varepsilon_1}+\frac{1}{\varepsilon_2}\right)\frac{e^2}{r} \tag{11}$$

In the exciton Hamiltonian (10), the potentials

$$V_h(r_h) = \begin{cases} 0, & r_h \le a \\ \infty, & r_h > a \end{cases},$$
$$V_e(r_e) = \infty, \quad r_e \le a \tag{12}$$

describe the quasiparticle motion using the models of an infinitely deep potential well.

As the QD radius *a* increases (so that *a* ≫*a* 0 *ex* ≫aex0), the spherical interface of the two media (QD-matrix) passes to the plane <semiconductor material with the permittivity>-matrix interface. In this case, the exciton with the spatially separated electron and hole (the hole moves within the semiconductor material and the electron lies in the borosilicate glass matrix) becomes two-dimensional [20-22].

The primary contribution to the potential energy of the Hamiltonian (10) describing exciton motion in a nanosystem containing a large-radius QD, *a* ≫*a* 0 *ex* ≫aex0 , is made by the electron-hole Coulomb interaction energy $V_{eh}(r)$ (11). The energy of the hole and electron interaction with self- $V_{hh'}(r_h, a, \varepsilon)$ (6), $V_{ee'}(r_h, a)$ (7) and "foreign" $V_{eh}'(\boldsymbol{r}_e, \boldsymbol{r}_h, a)$ (9), $V_{he'}(\boldsymbol{r}_e, \boldsymbol{r}_h, a)$ (8) images make a significantly smaller contribution to the potential energy of the Hamiltonian (10). In the first approximation, this contribution can be disregarded. In this case, only the electron-hole Coulomb interaction energy (11) remains in the potential energy of the Hamiltonian (10) [20-22]. The Schrodinger equation with such a Hamiltonian describes a two-dimensional exciton with a spatially separated electron and hole (the electron moves within the matrix, and the hole lies in the semiconductor material with the permittivity ε_2), the energy spectrum of which takes the following form [33, 34]:

$$E_n = -\frac{Ry^0_{ex}}{(n+1/2)^2},$$

$$Ry^0_{ex} = \frac{(\varepsilon_1+\varepsilon_2)^2}{4\varepsilon_1^2\varepsilon_2^2}\left(\frac{\mu_0}{m_0}\right)Ry_0 \qquad (13)$$

where n = 0, 1, 2... is the principal quantum number of the exciton and Ry_0 = 13.606 eV is the Rydberg constant. The Bohr radius of such a two-dimensional exciton is described by the following formula:

$$a^0_{ex} = \frac{2\varepsilon_1\varepsilon_2\hbar^2}{\varepsilon_1+\varepsilon_2\mu_0 e^2}, \qquad (14)$$

and the binding energy of the ground state of such a two-dimensional exciton, according to (13), is written as:

$$E^0_{ex} = -4Ry^0_{ex} \qquad (15)$$

The binding energy (15) of the exciton ground state is understood as the energy required for bound electron and hole state decay (in a state where n – 0).

To determine the ground-state energy of an exciton (with a spatially separated electron and hole) in a nanosystem containing QDs of the radius a, we applied the variational method. When choosing the variational exciton wave function, we used an approach similar to that developed in [14]. Let us write the variational radial wave function of the exciton ground-state (1*s* electron state and 1*s* hole state) in the nanosystem under study in the following form [20-22]:

$$\psi_0(\boldsymbol{r}_e,\boldsymbol{r}_h,r,a)=A\exp\left(-\frac{\mu(a)}{\mu_0}\frac{r}{a_{ex}^0}\right)\frac{(a^2-r_h^2)}{a^2}$$
$$\times\frac{(r_e^2-a^2)}{a^2}\frac{r}{a}\frac{\left|\boldsymbol{r}_e-(a/r_h)^2\boldsymbol{r}_h\right|r_h}{a^2}\frac{\left|\boldsymbol{r}_h-(a/r_e)^2\boldsymbol{r}_e\right|r_e}{a^2} \tag{16}$$

Here, the coefficient A is determined from the condition of normalization of the exciton wave function (16):

$$\int_a^{\infty} r_e dr_e \int_0^a r_h dr_h \int_r^{r_e+r_h} \psi_0^2(\boldsymbol{r}_e,\boldsymbol{r}_h,r,a)\,r dr=1$$

and the effectively reduced exciton mass $\mu(a)$ is the variational parameter.

As the QD radius a increases (so that $a \gg a\ 0\ ex \gg$aex0), a two-dimensional exciton is formed in the nanosystem. This leads to the variational exciton wave function (16) containing the Wannier-Mott two-dimensional exciton wave eigenfunction [33, 34]. Furthermore, the polynomials from r_e and r_h enter the exciton variational function (16), which make it possible to eliminate singularities in the functional $E_0(a,\mu(a))$ in the final analytical form.

To determine the exciton ground-state energy $E_0(a,\ \varepsilon)$ in the nanosystem under study using the variational method, we wrote the average value of the exciton Hamiltonian (10) in wave functions (16) as follows:

$$E_0(a,\mu(a))=\psi_0(\boldsymbol{r}_e,\boldsymbol{r}_h,r,a)\left|H(\boldsymbol{r}_e,\boldsymbol{r}_h,r,a)\right|\psi_0(\boldsymbol{r}_e,\boldsymbol{r}_h,r,a)=$$
$$\int_a^{\infty} dr_e \int_0^a dr_h \int_r^{r_e+r_h} dr\, r\, r_e r_h \psi_0(\boldsymbol{r}_e,\boldsymbol{r}_h,r,a)H(\boldsymbol{r}_e,\boldsymbol{r}_h,r,a)\psi_0(\boldsymbol{r}_e,\boldsymbol{r}_h,r,a) \tag{17}$$

The dependence of the energy $E_0(a)$ of the exciton ground state (n_e = 1, $l_e = m_e = 0$; $n_h = 1$, $l_h = m_h = 0$) (n_e, l_e, m_e and n_h, l_h, m_h are the principal, orbital and magnetic quantum numbers of the electron and hole, respectively) on the QD radius, a is calculated by minimizing the functional $E_0(a, \mu(a))$ (17):

$$\frac{\partial E_0\left(a, \mu(a)\right)}{\partial \mu(a)} = F\left(\mu(a), a\right) \tag{18}$$

Without writing cumbersome expressions for the first derivative of the functional $\partial E_0(a, \mu(a))/\partial\mu(a) = F(\mu(a), a)$, we present the numerical solution to the equation $F(\mu(a), a) = 0$ (18) in tabulated form. This follows from the table that the solution to this equation is the function $\mu(a)$, which monotonically varies weakly within the limits [20-22]:

$$0.304 \leq \mu(a) / m_0 \leq 0.359 \tag{19}$$

as the QD radius a varies within the range

$$2.0 \leq a \leq 29.8 \text{ nm} \tag{20}$$

(m_0 is the electron mass in a vacuum). In this case, the reduced exciton effective mass $\mu(a)$ (19) in the nanosystem differs slightly from the effective mass of an exciton (with a spatially separated hole and electron) $\mu = 0.304m_0$ by the value $(\mu(a) - \mu_0)/\mu_0 \leq 0.18$ when the QD radii vary within the range (20).

Simultaneously substituting the values of the variational parameter $\mu(a)$ (19) from Table 1 with the corresponding QD radii from the range (20) into the functional $E_0(a, \mu(a))$ (17), we obtain the exciton ground-state energy $E_0(a, \varepsilon)$ (17) as a function of the QD radius a [20-22].

Table 1. Variational parameter $\mu(a)$ as a function of the zinc selenide QD radius a.

a, nm	$\mu(a)/m_0$
2.0	0.359
3.0	0.352
4.0	0.345
5.0	0.338
6.0	0.331
8.0	0.325
10.0	0.319
15.0	0.313
20.0	0.308
29.8	0.304

The results of the variational calculation of the energy of the ground state of an exciton $E_0(a, \varepsilon)$ (17) in the nanosystem under study containing zinc selenide QDs of the radius a (20) are shown in the Figure 1 [20-22]. Here, the values of function $\mu(a)$ (19) and the results of the variational calculation of the exciton ground-state energy $E_0(a, \varepsilon)$ (17) are obtained for a nanosystem containing zinc selenide QDs, synthesized in a borosilicate glass matrix, as studied in the experimental works [10, 19].

In the experimental work [10], borosilicate glass samples doped with zinc selenide with concentrations ranging from x = 0.003-1%, obtained by the sol-gel method, were studied. According to X-ray diffraction measurements, the average radii a of ZnSe QDs formed in the samples were within $a \approx$ 2.0-4.8 nm. In this case, the values of $a^{\sim}$ a˜ were comparable to the exciton Bohr radius $a_{ex} \approx$ 3.7 nm in a zinc selenide single crystal. At low QD concentrations (x = 0.003 and 0.06%), their interaction can be disregarded. The optical properties of such nanosystems are primarily controlled by the energy spectra of electrons and holes localized near the spherical surface of individual QDs synthesized in the borosilicate glass matrix.

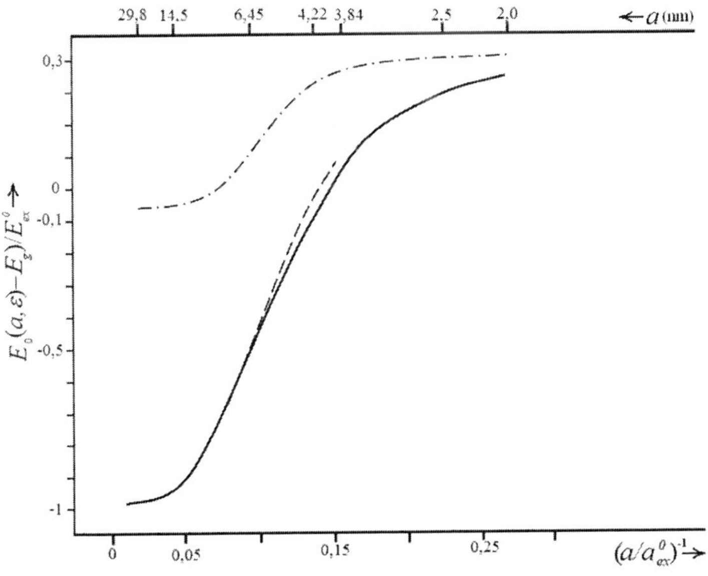

Figure 1. Dependences of the exciton ground state-$(E_0(a,\varepsilon) - E_g)$ (17) (solid curve) and the binding energy of the exciton ground state $(E_{ex}(a,\varepsilon) - E_g)$ (21) (dashed curve) on the zinc selenide QD radius a in the model of an exciton with a spatially separated electron and hole. The dash-dotted curve is the dependence of the exciton ground-state energy (E_0(a, ε) – E_g) on the zinc selenide QD radius a in the exciton model, in which the electron and hole move within the zinc selenide QD volume [16]. Eg = 2.823 eV is the band gap in a zinc selenide single crystal; *E* 0 *ex* E_{ex}^0 = 1.5296 eV (15) and *a* 0 *ex* aex0 , = 0.573 nm (14) are, respectively, the binding energy of the ground state and the Bohr radius of a two-dimensional exciton with a spatially separated electron and hole.

In [10, 19], a peak in the low-temperature luminescence spectrum at an energy of $E1 \approx 2.66$ eV was observed at the temperature $T = 4.5$ K in samples with $x = 0.06\%$; this energy is lower than the band gap of a zinc selenide single crystal (E_g = 2.823 eV). The shift of the peak of the low-temperature luminescence spectrum with respect to the band gap of the ZnSe single crystal to the short-wavelength region is $\Delta E1 = (E1 - E_g) \approx -165$ meV. The authors of [10] assumed that the shift $\Delta E1$ was caused by quantum confinement of the energy spectra of electrons and holes localized near the spherical surface of individual QDs, and that it was associated with a decrease in the average radii a of zinc-selenide QDs at low concentrations (x = 0.06%). In this case, the problem of the quantum confinement of which electron and hole states (the hole moving within the QD volume and the electron localized at the outer spherical QD-dielectric matrix interface or the electron and hole localized in the QD volume) caused such a shift of the luminescence-spectrum peak remained open.

Comparing the exciton ground-state energy $(E_0(a,\varepsilon) - E_g)$ (17) with the energy of the shift in the luminescence-spectrum peak $\Delta E1 \approx -165$ meV, we obtained the average zinc selenide QD radius $a1 \approx 4.22$ nm (see Figure 1) [20-22]. The QD radius $a1$ may be slightly overestimated, since the variational calculation of the exciton ground-state energy can yield slightly overestimated energies [33, 34]. The determined average QD radius $a1$ was found to be within the range of the average radii of zinc selenide QDs (a a $\approx$ 2.0-4.8 nm) studied under the experimental conditions of [10, 19].

It should be noted that the average Coulomb interaction energy $\bar{V}_{eh}(a, \varepsilon) = \psi_0(r_e, r_h, r, a) \mid V_{eh}(r) \mid \psi_0(r_e, r_h, r, a)$ between the electron and hole primarily contributed to the ground-state energy (17) of the exciton in the nanosystem containing zinc selenide QDs with radii $a1$ comparable to the exciton Bohr radius in a zinc-selenide single crystal ($a_{ex} \approx 3.7$ nm). In this case, the average energy of the interaction of the electron and hole with self- and "foreign" images,

$$\left(\bar{V}_{ee'}(a,\varepsilon) + \bar{V}_{hh'}(a,\varepsilon) + \bar{V}_{eh'}(a,\varepsilon) + \bar{V}_{he'}(a,\varepsilon) \right) = \\ = \langle \psi_0(r_e, r_h, r, a) \mid V_{ee'}(r_e, a, \varepsilon) + V_{hh'}(r_h, a, \varepsilon) + V_{eh'}(r_e, r_h, a, \varepsilon) + V_{he'}(r_e, r_h, a, \varepsilon) \mid \psi_0(r_e, r_h, r, a) \rangle$$

yielded a significantly smaller contribution to the exciton ground-state energy (17),

$$0.04 \le \left| \left[V_{ee'}(a, \varepsilon) + V_{hh'}(a, \varepsilon) + V_{eh'}(a, \varepsilon) + V_{he'}(a,\varepsilon) \right] / E_0(a, \varepsilon) \right| \le 0.12 \; [20-22].$$

Thus, the short-wavelength shift $\Delta E1$ of the low temperature luminescence spectrum peak is caused by renormalization of the electron-hole Coulomb interaction energy $V_{eh}(r)$ (11), as well as renormalization of the energy $U(r_e, r_h, r, a, \varepsilon)$ (5) of the polarization interaction of the electron and hole with the spherical QD-dielectric matrix interface, which is associated with spatial confinement of the quantization region by the QD volume. In this case, the hole moves within the QD volume and the electron is localized at the outer spherical QD-dielectric matrix interface.

The binding energy of the ground state of an exciton (with a spatially separated electron and hole) $E_{ex}(a, \varepsilon)$ in a nanosystem containing zinc selenide QDs of the radius a is the solution to the radial Schrodinger equation with a Hamiltonian containing, in contrast to Hamiltonian (10), only the terms $V_{he'}(r_e, r_h, a, \varepsilon)$ (8) and $V_{eh'}(r_e, r_h, a, \varepsilon)$ (9) in the polarization interaction energy $U(\mathbf{r}_e, \mathbf{r}_h, a, \varepsilon)$ (5), which describe the energies of the hole and electron interaction with "foreign" images, respectively [15, 27, 28]. Therefore, the exciton ground-state binding energy $Eex(a, \varepsilon)$ is defined by the expression [20-22]:

$$E_{ex}(a,\varepsilon)=E_0(a,\varepsilon)-\left\langle \psi_0(r_e,r_h,r,a)\middle|\left(V_{hh'}(r_h,a,\varepsilon)+V_{ee'}(r_e,a,\varepsilon)\right)\middle|\psi_0(r_e,r_h,r,a)\right\rangle, \tag{21}$$

where the term

$\langle\psi_0(r_e, r_h, r, a) \mid (V_{hh'}(r_h, a, \varepsilon)+V_{ee'}(r_e, a, \varepsilon)) \mid \psi_0(r_e, r_h, r, a)\rangle$ describes the average energy of hole and electron interaction with self-images.

Since the average energies of the interaction of the hole with its image and the average energies of the interaction of the electron with its image deliver contributions that take opposing signs to expression (21), they significantly compensate for each other. Therefore, the binding energies of the exciton ground state $E_{ex}(a, \varepsilon)$ (21) slightly differs from the corresponding total energies of the exciton ground state $E_0(a, \varepsilon)$ (17). This difference,

$$\Delta = \left| \left(E_{ex}(a,\varepsilon) - E_0(a,\varepsilon) \right) / E_{ex}(a,\varepsilon) \right|,$$

varies within $\Delta \leq 4\%$, as QD radii a varies within the range $3.84 \leq a \leq 8.2$ nm (see Figure 1) [20-22].

Figure 1 shows the dependences of the total energy $E_0(a, \varepsilon)$ (17) and the binding energy $E_{ex}(a, \varepsilon)$ (21) of the ground state of the exciton with a spatially separated electron and hole on the QD size for a nanosystem containing zinc selenide QDs of the radius a. We can see that the bound states of electron-hole pairs arise near the spherical surface of the QD, starting from the QD critical radius $a \geq a_c^{(1)} \approx 3.84$ nm. In this case, the hole is localized near the QD inner surface and the electron is localized at the outer spherical QD-dielectric matrix interface. Starting from the QD radius $a \geq a_c^{(1)}$, the electron-hole pair states are in the region of negative energies (counted from the top of the band gap E_g for a zinc selenide single crystal), which corresponds to the electron-hole bound state [20-22, 29-23]. In this case, the electron-hole Coulomb interaction energy $V_{eh}(r)$ (11) and the energy $U(r_e, r_h, r, a, \varepsilon)$ (5) of the polarization interaction of the electron and hole with the spherical QD-dielectric matrix interface dominate the energy of the quantum confinement of the electron and hole in the nanosystem under study.

The total energy $|E_0(a, \varepsilon)|$ (17) and the binding energy $|E_{ex}(a, \varepsilon)|$ (21) of the ground state of the exciton with a spatially separated electron and hole increases with QD radius a. In the range of radii

$$4.0 \leq a \leq 2.8 \text{nm} \tag{22}$$

the binding energy $|E_{ex}(a, \varepsilon)|$ (21) of the exciton ground state significantly (by a factor of 4.1-76.2) exceeds the exciton binding energy in a zinc selenide single crystal, $E\,0\,ex$ ⁻⁻⁻⁻⁻ Eex0⁻ ≈ −21.07 meV. Starting from the QD radius $a \geq a_c^{(2)} \approx 29.8$ nm, the total energies (17) and binding energies (21) of the exciton ground state asymptotically tend to the value $E\,0\,ex$ Eex0 = −1.5296 eV, which characterizes the binding energy of the ground state of a two-dimensional exciton with a spatially separated electron and hole (see the figure 1) [20-22, 29-32].

The obtained values of the total energy $E_0(a, \varepsilon)$ (17) of the exciton ground state in the nanosystem satisfy the inequality

$$\left(E_0(a,\varepsilon) \quad E_g\right) \ll \Delta V(a) \tag{23}$$

where $\Delta V(a)$ is the potential-well depth for the QD electron. For a large class of II-VI semiconductors in the region of QD sizes, $a \geq a_{ex}^0$ aex0 , $\Delta V(a)$ = 2.3-2.5 eV [7]. Satisfaction of condition (23) likely makes it possible to disregard the effect of the complex structure of the QD valence band on the total energy (17) and the binding energy (21) of the exciton ground state in the nanosystem under study when deriving these quantities.

The effect of a significant increase in the binding energy $|E_{ex}(a, \varepsilon)|$ (21) of the exciton ground state in the nanosystem under study, according to formulas (5) to (9), (11), (13) to (15), (17) and (21) is controlled by two factors [20-22, 29-32]: (i) a significant increase in the energy of the electron-hole Coulomb interaction $|V_{eh}(r)|$ (11) and an increase in the energy of the interaction of the electron and hole with "foreign" images $|V_{eh'}(r_e, r_h, r, a, \varepsilon)|$ (9), $|V_{he'}(r_e, r_h, r, a, \varepsilon)|$ (8) (the "dielectric enhancement" effect [34]); (ii) spatial confinement of the quantization region by the QD volume. In this case, as the QD radius a increases, starting from $a \geq a_c^{(2)}$ ≥ac(2) $\approx 52\, a_{ex}^0$ 52 aex0 $\approx$ 29.8 nm, the exciton becomes two-dimensional, with a ground-state energy of E_{ex}^0 Eex0 (15), which exceeds the exciton binding energy E_{ex} in the zinc selenide single crystal by almost two orders of magnitude:

$$\left(\left|E_{ex}^0 / \overline{E_{ex}^0}\right| \approx 72.6\right).$$

The "dielectric enhancement" effect is caused by the following factor. When the matrix permittivity ε_1 is significantly smaller than the QD permittivity ε_2, the most important role in the electron-hole interaction in the nanosystem under study is fulfilled by the field induced by these quasiparticles in the matrix. In this case, electron-hole interaction in the nanosystem appears to be significantly stronger than in an infinite semiconductor with the permittivity ε_2 [34].

In [16], in the nanosystem experimentally studied in [10], an exciton model in which the electron and hole move within the zinc selenide QD volume was studied. Using the variational method, within the modified effective mass method, the dependence of the exciton ground-state energy $E_0(a, \varepsilon)$ on the QD radius a in the range (20) was obtained in [16] (see Figure 1). It was shown that, as the QD radius increased, starting from $a \geq a_c = 3.90\ a\ 0\ ex$ ----- aex0‾ = 1.45 nm, a bulk exciton appeared in the QD; its binding energy,

$$\overline{E_{ex}^{0}} = -\frac{\hbar^2}{2\mu\left(\overline{a_{ex}^{0}}\right)^2} \tag{24}$$

was 21.07 me $(\mu = 0.132 m_0 \text{ and } a_{ex}^0 = 3.7$ nm are the reduced effective mass and Bohr radius of the exciton in the zinc-selenide forming the QD volume). The bulk exciton in the QD is understood as an exciton whose structure (reduced mass, Bohr radius and binding energy) in the QD does not differ from the structure of an exciton in an infinite semiconductor material. As the QD radius a increases ($a \geq a_c$), the exciton ground-state energy $E_0(a)$ asymptotically follows the binding energy of the bulk exciton (24) (see Figure 1) [20-22, 29-32].

Thus, using the exciton model, in which an electron and hole move in the QD volume, it is impossible to interpret the mechanism of the appearance of the nanosystem luminescence-spectrum peak with the shift $\Delta E1 \approx -165$ meV, obtained in [10, 19].

A comparison of the dependences of the exciton ground-state energy $E_0(a)$ in the nanosystem [10], obtained using two-exciton models (see Figure 1) (the electron and hole move within the zinc selenide QD volume [16]) (model **I**); the hole moves within the zinc selenide QD volume and the electron is localized in the boron silicate glass matrix near the QD spherical surface (model **II**), allowing for the following conclusion. In model **I**, as the QD radius a increases, starting from $a \geq a_c \approx 14.5$ nm, the exciton ground- state energy $E_0(a)$ asymptotically follows the binding energy of the bulk exciton E_{ex}^{0} Eex0ˉ ≈ –21.07 meV (24). In model II, as the QD radius increases, starting from $a \geq a_c^{(2)}$ ac(2) ≈ 29.8 nm, the exciton ground-state energy (17) asymptotically follows E_{ex}^{0} Eex0 = – 1.5296 eV (15) (characterizing the binding energy of the ground state of a two- dimensional exciton with a spatially separated electron and hole), which is significantly lower than E_{ex}^{0} Eex0ˉ ≈ –21.07 meV [20-22, 29-32].

3. Excitonic quasimolecules formed from spatially separated electrons and holes

We considered a model nanosystem [23, 24] that consisted of two spherical semiconductor QDs, A and B, synthesized in a borosilicate glass matrix with the permittivity ε_1 ε1 . Let the QD radii be a and the spacing between the spherical QD surfaces be D. Each QD is formed from a semiconductor material with the permittivity ε_2 ε2 . For simplicity, without loss of generality, we assumed that holes h (A) and h (B) with the effective masses m_h were in the QD (A) and QD (B) centres and the electrons $e(1)$ and $e(2)$ with the effective masses $m_e^{(1)}$ me(1) were localized near the spherical QD(A) and QD (B) surfaces, respectively. The above assumption was reasonable, since the ratio between the effective masses of the electron and hole in the nanosystem was much smaller that unity: $((m_e^{(1)}/m_h) \ll 1)$ ((me(1)/mh)≪1) . Let us assume that there was an infinitely high potential barrier at the spherical QD-matrix interface. Therefore, in the nanosystem, holes did not leave the QD bulk and electrons did not penetrate into the QDs.

In the context of the adiabatic approximation and effective mass approximation, using the variational method, we obtained the total energy $E_0(\tilde{D},\tilde{a})$ E0(D˜,a˜) and the binding energy $E_e(\tilde{D},\tilde{a})$ Ee(D˜,a˜) of the biexciton singlet ground state (the spinning of the electrons $e(1)$ and $e(2)$ were antiparallel) in such a system as functions of the spacing between the QD surfaces D and the QD radius a [23, 24]:

$$E_0\left(\tilde{D},\tilde{a}\right)=2E_{ex}\left(\tilde{a}\right)+E\left(\tilde{D},\tilde{a}\right),$$

(24)

Here, the binding energy $E_{ex}(\tilde{a})$ Eex(a˜) (17) of the ground state of the exciton (formed from an electron and a hole spatially separated from the electron) localized above the QD(A) (or QD(B)) surface was determined in [23, 24] (by the parameters $\tilde{a}=(a/a_{ex}^0)$ a˜=(a/aex0) ($a_{ex}^0=3$ aex0= 3 .7 nm and the exciton Bohr radius in a single crystal ZnSe, $\tilde{D}=(D/a_{ex}^0)$ D˜=(D/aex0)). For the nanosystem under study, the values of the binding energies $E_{ex}(\tilde{a})$ Eex(a˜) were calculated in [23, 24] for use in the experimental conditions of [10, 19].

The results of the variational calculation of the binding energy $E_{в}(\tilde{D},\tilde{a})$ Eв(D˜,a˜) of the biexciton singlet ground state in the nanosystem of ZnSe and QDs with an average radii of $\bar{a}_1=3.88$ nm a¯1=3.88 nm „ synthesized in a borosilicate glass matrix, are shown in [23, 24]. Such a nanosystem was experimentally studied in [10, 19]. In [10, 19], the borosilicate glassy samples doped with ZnSe to the content x from x = 0.003-1% were produced using the sol-gel technique. At a QD content of x = 0.06 %, one must take into account the interaction of charge carriers localized above the QD surfaces.

The binding energy $E_{в}(\tilde{D},\tilde{a})$ Eв(D˜,a˜) of the biexciton singlet ground state in the nanosystem of ZnSe QDs with average radii of $\bar{a}_1=3.88$ nm a¯1=3.88 nm has a minimum of $E_{в}^{(1)}(D_1,\bar{a}_1)\approx-4,2$ meV Eв(1)(D1, a¯1)≈−4,2 meV (at the spacing $D_1\cong3.2$ D1≅3.2 nm) [23, 24]. The value of $E_{в}^{(1)}$ corresponds to the temperature $T_c\approx49$ K. Tc≈49 K. In [23, 24], it follows that a biexciton (excitonic quasimolecule) is formed in the nanosystem, starting from a spacing between the QD surfaces of $D\geq D_c^{(1)}\cong2,4$ nm. The formation of such a excitonic quasimolecule (biexciton) is of the threshold character and possible only in a nanosystem with QDs with average radii $\bar{a}_1$, a¯1, such that the spacing between the QD surfaces D

exceeds a certain critical spacing $D_c^{(1)}$ Dc(1) . Moreover, the exciton quasimolecule (biexciton) can exist only at temperatures below a certain critical temperature, i.e., $T_c \approx 49$ K Tc≈49 K [23, 24].

As follows from the results of variational calculation [23, 24], the binding energy of an exciton (formed from an electron and a hole spatially separated from the electron) localized above the surface of the QD(A) (or a QD(B)) with an average radius of $\tilde{a}_1$ =3,88 nm a˜1=3,88 nm is $E_{ex}(\tilde{a}_1) \cong -54$ meV . In this case, the energy of the biexciton singlet ground state $E_0(\tilde{D}, \tilde{a})$ (25) takes the value $E_0(\tilde{D}, \tilde{a}) = -112$meV.

From the results of variational calculation [23, 24], of the biexciton (exciton quasimolecule) binding energy $E_e(\tilde{D}, \tilde{a})$, it follows that the major contribution to the binding energy (25) is made by the average energy of the exchange interaction of the electrons $e(1)$ and $e(2)$ alongside holes $h\ (A)$ and $h\ (B)$. At the same time, the energy of the Coulomb interaction makes a much smaller contribution of the biexciton binding energy $E_e(\tilde{D}, \tilde{a})$ Ee(D˜,a˜) (25).

The major contribution to the exchange is interaction energy, created by the energy of the exchange interaction of the electron $e(1)$ with the holes $h\ (B)$, as well as of the electron $e(2)$ with the holes $h\ (B)$, and of the electron $e(2)$ with the holes $h\ (A)$. The major contribution to the Coulomb is interaction energy, created by the energy of the Coulomb interaction of the electron $e(1)$ with the holes $h\ (B)$, as well as of the electron $e(2)$ with the holes $h\ (A)$ [23, 24].

As the spacing D between the QD(A) and QD(B) surfaces is increased, starting from $D \geq D_c^{(2)} \cong 16{,}4$ nm D≥Dc(2)≅16,4 nm , the average Coulomb interaction energy substantially decreases. In addition, because of the decrease in the overlapping of the electron wave function, the average exchange interaction energy also substantially decreases. Consequently, the average Coulomb interaction energy and the average energy of the exchange interaction of the electrons $e(1)$ and $e(2)$ with the holes $h\ (A)$ and $h\ (B)$ sharply decrease in comparison with the exciton binding energy $E_{ex}(\tilde{a})$ Eex(a˜) (17) [23, 24], resulting in decomposition of the biexciton in the nanosystem into two excitons (formed of spatially separated electrons and holes) localized above the QD(A) and QD(B) surfaces.

4. THEORY OF NEW SUPERATOMS — ANALOGUE ATOMS FROM THE GROUP OF ALKALI METALS

The idea of superatoms (or artificial atoms) is essential for the development of mesoscopic physics and chemistry [20-22, 29, 30]. Superatoms are nanosized quasi-atomic nanostructures formed from spatially separated electrons and holes (the hole in the volume of the QD and the electron is localized on the outer spherical quantum dot matrix dielectric interface) [20-22, 29, 30]. This terminology can be accepted as correct, given the similarities between the spectra of discrete electronic states of atoms and superatomic atoms, and the similarities in terms of their chemical activities [20-22, 29, 30].

In [20-22], within the framework of the modified effective mass method [14], the theory of artificial atoms formed from spatially separated electrons and holes (holes moving in the volume of a semiconductor (dielectric) QD and an electron localized on the outer spherical interface between the QD and a dielectric matrix) is developed. The energy spectrum of superatoms (excitons of spatially separated electrons and holes) from QD radius $a \geq ac$ (about 4 nm) is fully discrete [20-22, 29, 30]. This is referred to as a hydrogen-superatom and is localized on the surface of a valence electron QD. The energy spectrum of the superatom consists of a quantum-dimension of discrete energy levels in the band gap of the dielectric matrix. Electrons in superatoms are localized in the vicinity of the nucleus (QD). The electrons move in well-defined atomic orbitals and serve as the nucleus of QD, containing in its volume semiconductors and insulators. Ionization energy superatoms take on large values (of the order of 2.5 eV), which is almost three orders of magnitude higher than the binding energy of the excitons in semiconductors [20-22, 29, 30].

We will briefly discuss the possible physical and chemical effects that are relevant for the results of this paper. In our proposed [20-22, 29, 30] model of a hydrogen superatom localized on the surface of the

QD is a valence electron. In quasi-atomic structures of the outer valence, electrons can participate in a variety of physical and chemical processes, similar to the atomic valence electrons in atomic structures. Artificial atoms have the ability to connect to the electron orbitals of electrons *N* (where *N* can vary from one to several tens). At the same time, the number of electrons *N* can take values of the order of a few tens or even surpass the serial numbers of all the known elements found in Mendeleev's table [20-22, 29, 30]. This new effect allows for attaching to the electronic orbitals of artificial atoms *N* electrons, causing a high reactivity and opening up new possibilities for superatoms related to their strong oxidizing properties, increasing the possibility of substantial intensity in photochemical reactions during catalysis and adsorption, as well as their ability to form many new compounds with unique properties (in particular, the quasi-molecule and quasicrystals) [24, 29, 30]. Therefore, studies aimed at the theoretical prediction of the possible existence of artificial new atoms (not listed in the Mendeleev table) and to their study in terms of experimental conditions are very relevant.

Quantum discrete states of the individual atoms of alkali metals are determined by the movement of only one, i.e., the outermost valence electron, around a symmetric atomic core (containing the nucleus and the remaining electrons) [35]. In the hydrogen superatom formed quantum-energy spectra of discrete energy levels of the valence electron [20-22, 29, 30]. Thus, the observed similarity of the spectra of discrete electronic states and individual superatoms alkali metal atoms, as well as the similarity of their chemical activity [20-22, 29, 30, 35].

In Section 4, on the basis of a spectroscopic analogy of electronic states of artificial atoms and individual alkali metal atoms, a new artificial atom is theoretically predicted, which is similar to the new alkali metal atom.

In [20-22, 29, 30], a new model of an superatom is proposed, which is a quasi-zero-dimensional nanosystem consisting of a spherical QD (nucleus superatom) with radius *a* and which is included within its scope as a semiconductor (dielectric) with a dielectric constant ε_2, surrounded by a dielectric matrix with a dielectric constant ε_1. A hole h with the effective mass m_h moves in the QD volume, while an electron e with the effective mass $m_e^{(1)}$ lies in the dielectric matrix. In such a nanostructure, the lowest electronic level is situated in the matrix and the humble hole level is the volume QD. Large shift of the valence band (about 700 meV) is the localization of holes in the volume QD. A large shift of the conduction band (about 400 meV) is a potential barrier for electrons (electrons move in the matrix and do not penetrate into the volume QD). The Coulomb interaction energy of an electron and a hole, and the energy of the electron polarization interaction with the surface section (QD-matrix) (since the permittivity ε_2 is far superior to QD permittivity ε_1 matrix) cause localization of the electron in the potential well above the surface of QD [20-22, 29, 30].

With increasing radius a QD, so that $a \gg a_{ex}^{0}$ (where a_{ex}^{0} (14) two-dimensional Bohr radius of the electron) spherical surface section (QD- matrix) transforms into a flat surface section. In this artificial atom, electrons localized on the surface (QD-matrix) become two-dimensional. In this case, the potential energy in the Hamiltonian describing the motion of an electron in a superatom, the main contribution to the energy of the Coulomb interaction $V_{eh}(r)$ (11) between an electron and a hole [20-22]. Polarization interaction energy of the electron and the hole with a spherical surface section (QD-matrix) delivers a much smaller contribution to the potential energy of the Hamiltonian and thus, contributions to a first approximation can be neglected [20-22]. In this regard, the two-dimensional electron energy spectrum En in the artificial atom takes the form (13).

Depending on the binding energy $E_{ex}(a,\varepsilon)$ of an electron in the ground state superatom (QD containing zinc-selenide radius a and surrounded by a matrix of borosilicate glass [10]) as obtained in [20-22] by the variational method, it follows that the bound state of an electron occurs near the spherical interface (QD-matrix), starting with the value of the critical radius QD $a \geq a_c^{(1)} = 3.84$ nm, when this hole moves in the volume QD and the electron is localized on the surface of the spherical section (QD-matrix). In this case, the

Coulomb interaction energy $V_{eh}(r)$ (11) between the electron and the hole, and the energy of the polarization interaction of electrons and holes with a spherical surface section (QD-matrix) prevail over the size quantization of the energy of electrons and holes in the artificial atom. Thus, [20-22] found that the occurrence of superatoms had a threshold and was only possible if the radius of QD KT $a \geq a_c^{(1)} = 3.84$ nm.

With the increasing radius of *a* QD scan, an increase in the binding energy of the electron in the ground state superatom was observed. In the range of radii $4.0 \leq a \leq 29.8$ nm, the binding energy of the electron in the ground state superatom significantly exceeded (in (4,1-76,2) times) the value of the exciton binding energy $\tilde{E}^0_{ex} \approx 21.07$ meV in a single crystal of zinc-selenide [20-22]. Beginning with *a* radius QD $a \geq a_c^{(2)} = 29.8$ nm, the energy of the ground state of an electron in a superatom asymptotically follow the value $E^0_{ex} = -1.5296$ eV, which characterized the energy of the ground state of two-dimensional electrons in an artificial atom (15) [20-22].

The effect of significantly increasing the energy of the ground state of an electron in a superatom was primarily determined by two factors [20-22]: 1) a significant increase in the Coulomb interaction energy $|V_{eh}(r)|$ (2) electron-hole (the "dielectric enhancement" [34]); 2) the spatial limitations on the quantization volume QD, while with an increasing radius of *a* QD, since the radius of QD $a \geq a_c^{(2)} = 52a^0_{ex} = 29.8$ nm, superatoms became two-dimensional with a binding energy of the ground state E^0_{ex} (15), the value of which exceeded the exciton binding energy in a single crystal of zinc-selenide by two orders. The effect of "dielectric enhancement" as a result of the dielectric constant ε_1 of the matrix was much lower than the dielectric constant of QD ε_2, which played an essential role in the interaction between the electron and the hole in the superatom playing field produced by these quasi-particles in a matrix. Thus, the interaction between the electron and the hole in the superatom was significantly larger than in a semiconductor permittivity ε_2 [34].

Quantum discrete states of the individual atoms of alkali metals were determined by the movement of only one, the outermost valence electron, around a symmetric atomic core (containing the nucleus and the remaining electrons) [35]. Where large distances were the case between *r* electron and the nucleus (so that $r >> a0$, where $a0$ = 0.053 nm – the Bohr radius of the electron in a hydrogen atom), the field of the atomic core was described by the Coulomb field [35]:

$$V(r) = -\left(Ze^2 / r\right), \tag{26}$$

determining the interaction of the valence electron with the atomic core (Z – serial number of the atom in the periodic table of Mendeleev). The energy spectrum of a single atom of an alkali metal hydrogen-described spectrum [35] is given as follows:

$$E_{n^*} = -\frac{Ry^*}{\left(n^*\right)^2}, \; Ry^* = Z^2\, Ry_0 \tag{27}$$

where $n^* = (n + y)$ and effective quantum number (n = 1, 2, 3,... the principal quantum number); the amendment y depended on the orbital quantum number l. Amendments to y were due to the fact that the valence electron moved in the Coulomb field of the atomic core, where the nuclear charge was screened by core electrons. Amendment y corrections were determined by comparing the spectrum of (6) with its experimental values. The value of $y < 0$ and was numerically closer to the atomic core suitable valence electron orbit. The number of possible orbits of the valence electron in a single alkali metal atom such as a hydrogen atom, and [35].

The similarity of the individual series of neutral alkali metal atoms with the hydrogen Balmer series suggests that the energy spectra of neutral alkali metal atoms can be labelled valence electron radiation in transition from higher levels to the level of principal quantum number $n = 2$ [35].

In a single atom of an alkali metal valence electron moving in the Coulomb field of the atomic core (26) having the same functional dependence on r as the Coulomb field (11), in which the valence electron in hydrogen-like model of artificial atom. This leads to the fact that the energy spectra of the valence electron in a single atom of an alkali metal (27) and in the artificial atom (13) describe the spectrum of hydrogen-type. At the same time, the number of possible quantum states of valence electrons in a hydrogen-like artificial atom model is the same as the number of quantum states of discrete valence electrons in a single atom of an alkali metal [20-22, 29, 30].

Table 2 shows the position of the valence electron energy levels in the atoms of alkali metals (K, Rb, Sc) [35] and the new artificial atom X, as well as the level shifts of the valence electron (*ΔE* κ *Rb* ΔERbκ , *ΔE Rb Sc* ΔEScRb , *ΔE Sc x* ΔExSc) relative to the adjacent level. Assume that the shift of the energy level E_x artificial atom X (relative to the energy level E_{sc} of the atom Sc) will be the same as the shift of the energy level E_{Rb} of the atom Rb (relative energy level E_{sc} of the atom Sc), (i.e., ΔE_x^{Sc} ΔE_{Sc}^{Rb}). . In this case, the level of the valence electron artificial atom will be *Ex* = –593 meV. Using the dependence of the binding energy $E_{ex}(a,\varepsilon)$ of the ground state of an electron in an artificial atom [20-22] (QD containing zinc-selenide radius *a* and surrounded by a matrix of borosilicate glass [10]), we found the radius QD zinc-selenide *a*1 = 5.4 nm, which corresponded to the *Ex* = – 593 meV. It should be noted that the energy levels of a valence electron in the individual atoms of alkali metals (K, Rb, Sc) [35] and the new artificial atom X are located in the infrared spectrum.

Table 2. Position of energy levels of the valence electron in some alkali metal atoms (K, Rb, Sc) and a new artificial atom, X. Level shifts of the valence electron (*ΔE* κ *Rb* ΔERbκ , *ΔE Rb Sc* ΔEScRb , *ΔE Sc x* ΔExSc) are relative to the adjacent level.

Alkali metal atoms selected	Valence electron energy levels (meV)	Level shifts of the valence electron (meV)
K	- 7 21.1	
Rb	- 7 11.2	1 0
Sc	- 652	5 9
X	- 5 93	5 9

Thus, we propose a new model of an artificial atom that is a quasi-atomic heterostructure consisting of a spherical QD (nucleus superatom) radius a and which contains in its scope, zinc- selenide, surrounded by a matrix of borosilicate glass (in volume QD, h hole effectively moves mass m_h, e and the electron effective mass $m_e^{(1)}$ is located in the matrix), thus allowing for finding a new artificial atom X (absent in the Mendeleev periodic system), which is similar to a new single alkali metal atom. This new artificial atom of a valence electron can participate in various physical [20-22, 29, 30] and chemical [30, 35] processes that are analogous to atomic valence electrons in atomic systems (in particular, the selected alkali metal atoms [35]). Such processes are unique as a result of the new properties of artificial atoms: strong oxidizing properties that increases the possibility of substantial intensity in photochemical reactions during catalysis and adsorption, as well as their ability to form a plurality of the novel compounds with unique properties (in particular, the quasi-molecule and the quasicrystals [23, 24]).

The application of semiconductor nanoheterostructures as the active region nanolasers prevents small exciton binding energy in QD. Therefore, studies aimed at finding nanoheterostructures, which will yield a significant increase in the binding energy of the local electronic states in QDs, are relevant [20-22]. The effect of significantly increasing the energy of the electron in a hydrogen superatom [20-22, 29, 30] allows for better experimental detection of the existence of such superatoms at room temperatures and will stimulate experimental studies of nanoheterostructures containing superatoms, which can be used as active region nanolasers when working with optical transitions.

5. CONCLUSIONS

The theory of an exciton with a spatially separated electron and hole was developed within the framework of the modified effective mass method [14], in which the reduced effective exciton mass is a function of the semiconductor QD radius a. The average zinc-selenide QD radius was determined by comparing the dependence of the exciton ground-state energy (17) on the QD radius, obtained by the variational method within the modified effective mass method [14] and using the experimental peak of the low-temperature luminescence spectrum [10, 19]. It was shown that the short-wavelength shift of the peak of the low-temperature luminescence spectrum of the samples containing zinc selenide QDs, which was observed under the experimental conditions noted in [10, 19], was caused by renormalization of the electron-hole Coulomb interaction energy (11), as well as the energy created by the polarization interaction (5) of the electron and hole with the spherical QD-dielectric matrix interface, related to spatial confinement of the quantization region by the QD volume. In this case, the hole moves in the QD volume and the electron is localized at the outer spherical QD-dielectric matrix interface [20-22, 29-32].

To apply semiconductor nanosystems containing zinc-selenide QDs as the active region of lasers, it is required that the exciton binding energy $|E_{ex}(a, \varepsilon)|$ (21) in the nanosystem be at the order of several kT_0 and at room temperature T_0 (k is the Boltzmann constant) [13]. Nanosystems consisting of zinc-selenide QDs grown in a borosilicate glass matrix can be used as the active region of semiconductor QD lasers. In the range of zinc selenide QD radii a (22), the parameter $|E_{ex}(a, \varepsilon)/kT_0|$ take on significant values ranging from 3.1 to 56 [20-22, 29-32].

The effect of significantly increasing the binding energy (21) of the exciton ground state in a nanosystem containing zinc selenide QDs with radii a (22) was detected; compared to the exciton binding energy in a zinc selenide single crystal, the increase factor was 4.1-72.6. [20-22, 29-32]. It was shown that the effect of significantly increasing the binding energy (21) of the exciton ground state in the nanosystem under study was controlled by two factors [20-22, 29-32]: (i) a substantial increase in the electron-hole Coulomb interaction energy (11) and an increase in the energy of the interaction of the electron and hole with "foreign" images (8), (9) (the "dielectric enhancement" effect [34]); (ii) spatial confinement of the quantization region by the QD volume; in this case, as the QD radius a increased, starting from $a \geq a$ (2) c ac(2) ≈ 29.8 nm, the exciton became two-dimensional with a ground-state energy (15), which exceeded the exciton binding energy in a zinc selenide single crystal by almost two orders of magnitude.

A review devoted to the theory of excitonic quasimolecules (biexciton) (made up of spatially separated electrons and holes) in a nanosystem that consists of ZnSe QDs synthesized in a borosilicate glass matrix was developed within the context of the modified effective mass approximation. Using the variational method, we obtained the total energy and the binding energy of the biexciton singlet ground state in such a system as functions of the spacing between the QD surfaces and the QD radius. It was established that, in a nanosystem composed of ZnSe QDs with the average radii $\bar{a}$ 1 $\bar{a}$1 , the formation of a biexciton (exciton quasimolecule) was of the threshold character and possible in a nanosystem where the spacing D between the QD surfaces is defined by the condition D (1) $c \leq D \leq D$ (2) c Dc(1)≤D≤Dc(2) [23, 24]. Moreover, the exciton quasimolecule (biexciton) can exist only at temperatures below a certain critical temperature, i.e., $T\,c \approx 49$ K Tc≈49 K [23, 24]. It was established that the spectral shift of the low temperature luminescence peak [10, 19] in such a nanosystem resulted due to quantum confinement of the energy of the biexciton singlet ground state.

Thus, we propose a new model of an artificial atom, which is a quasi-atomic heterostructure consisting of a spherical QD (nucleus superatom) with radius a and which contains in its scope zinc selenide, surrounded by a matrix of borosilicate glass (in volume QD moves h hole effective mass m_h, e and the electron effective mass $m_e^{(1)}$ is located in the matrix), and which is allowed to find a new artificial atom X (absent in the Mendeleev periodic system), which is similar to a new single alkali metal atom. This new artificial atom of valence electron can participate in various physical [20-22, 29, 30] and chemical [30, 35] processes that are analogous to atomic valence electrons in atomic systems (in particular, the selected alkali metal atoms [35]). Such processes are unique due to the new properties of artificial atoms: strong oxidizing properties that increase the possibility of substantial intensity in photochemical reactions during catalysis and adsorption, as well as their ability to form plurality among novel compounds with unique properties (in particular, the quasi-molecule and the quasicrystals [23, 24]).

REFERENCES

1. Ekimov, A., Onushchenko, A., (1981). Size quantization of electrons in microcrystals. *Journal Experimental Theoretical Physics Letters*. 34, 345-348.
2. Ekimov, A., Onushchenko, A., (1984). Size quantization of electrons in semiconductor microcrystals. *Journal Experimental Theoretical Physics Letters*. 40, 1136-1139.
3. Ekimov, A., Efros, A., (1985). Excitons in semiconductor microcrystals. *Solid State Communication*. 56, 921-924.
4. Ekimov, A., Onushchenko, A., Efros, A., (1986). Size quantization of electron-hole pairs in nanocrystals. *Journal Experimental Theoretical Physics Letters*. 43, 376-379.
5. Chepik, D., Efros, A., Ekimov, A., (1990). Spectroscopy of excitons in semiconductor nanocrystals. *Journal of Luminescence*. 47, 113-118.
6. Ekimov, A., Hache, F., Schanne-Klein, M., (2003). Optical properties semiconductor quantum dots. *Journal Optical Society American*. B 20, 100 -108.
7. Grabovskis, V., Dzenis, Y., Ekimov, A., (1989). Photoluminescences of excitons in semiconductor nanocrystals. *Physics Solid State*. 31, 149 -152.
8. Alferov, J.I., (2002). Progress development in semiconductor nanostructures. *Physics Uspekhi*. 172, 1068 - 1074.
9. Alferov, J.I., (1998). Optical properties in semiconductor nanostructures. *Semiconductors*. 32, 1-8.

10. Bondar, V.N., Brodin, M.S., (2010). Optical properties semiconductor quantum dots. *Semiconductors.* 44, 884-890.

11. Efros, A.L., Efros, Al. L., Interband absorption light in semiconductor sphere. (1982). *Sov. Phys. Semiconductors.* 16, 955- 962.

12. Pokutnyi, S.I., (2004). Size quantization Stark effect in semiconductor quantum dots. *J. Appl. Phys.* 96, 1115-1122.

13. Pokutnyi, S.I., (2005). Optical nanolaser heavy hole transitions in quasi-zero- dimensional semiconductors nanosystems. *Physics Lett. A.* 342, 347-352.

14. Pokutnyi, S.I., (2007). Exciton states in semiconductor quantum dots in the framework of the modified effective mass method. *Semiconductors.* 41, 1323-1331.

15. Pokutnyi, S.I., (2010). Exciton states in semiconductor quantum dots. *Semiconductors.* 44, 488-493.

16. Pokutnyi, S.I., (2012). Exciton states in semiconductor nanosystems. *Semiconductors.* 46, 174-184.

17. Soloviev, V., Eeichofer, A., (2001). Approximation of effective mass in nanosystems. *Physica Status Solidi B.* 224, 285-291.

18. Yeh, C., Zhang, S., Zunger, A., (2004). The effective mass method in nanosystems. *Physical Review B.* 62, 14408-14416.

19. Bondar, N., Brodyn, M., (2010). Spectroscopy of semiconductor quantum dots. *Physics E.* 42, 1549-1555.

20. Pokutnyi, S.I., (2013). Binding energy of the exciton of a spatially separated electron and hole in quasi-zero-dimensional semiconductor nanosystems. *Technical Physics Letters.* 39, 233-235.

21. Pokutnyi, S.I., (2013). On an exciton with a spatially separated electron and hole in quasi-zero-dimensional semiconductor nanosystems. *Semiconductors.* 47, 791 -798.

22. Pokutnyi, S.I., (2014). Theory of excitons formed from spatially separated electrons and holes in quasi-zero-dimensional semiconductor nanosystems. *SOP Transactions Theoretical Physics.* 1, 24 -35.

23. Pokutnyi, S.I., (2013). Biexcitons formed from spatially separated electrons and holes in quasi-zero-dimensional semiconductor nanosystems. *Semiconductors.* 47, 1626-1635.

24. Pokutnyi, S.I., (2014). Quasi-zero-dimensional nanostructures: Excitonic quasimolecules. *J. Appl. Chem.* 2, 1-4.

25. Efremov, N.A., Pokutny, S.I., (1985). Macroscopic local charge states in ultradispersion media. *Sov. Phys. Solid. State.* 27, 27-35.

26. Efremov, N.A., Pokutny, S.I., (1990). Energy spectrum of exciton in spherical particle. *Sov. Phys. Solid. State.* 32, 955-964.

27. Pokutnyi, S.I., (1992). Size quantization of electron-hole pair in semiconductor quantum dots. *Physics Lett. A.* 168, 433-438.

28. Pokutnyi, S.I., (2011). Theory of exciton states in semiconductor nanosystems. *Physics Express.* 1, 158-164.

29. Pokutnyi, S.I., Gorbyk, P.P., (2013). Superatoms in quasi-zero-dimensional nanostructures (review). *Progr. Phys. Metal.* 14, 144 - 168.

30. Pokutnyi, S.I., Gorbyk, P.P., (2013). Superatoms in quasi-zero-dimensional nanosystems. *J. Appl. Chem.* 1, 44 - 47.

31. Pokutnyi, S.I., (2012). Exciton states in quasi-zero-dimensional semiconductor nanosystems: Theory. *Phys. Express.* 2, 20 - 26.

32. Pokutnyi, S.I., (2013). Exciton states in quasi-zero-dimensional nanostructures. *J. Appl. Chem.* 1, 5 - 18.

33. Lozovik, Y., Nishanov, V., (1976). Exciton Wannier-Mott near the interface. *Physics Solid State.* 18, 1905-1911.

34. Keldysh, L., (1979). The interaction of the electrons in a thin films. *Journal Experimental Theoretical Physics Letters.* 29, 621-624.

35. Frish, S.E., (1963). Optical spectra of atoms. Nauka, Moscow [in Russian].

2

CHAPTER

WATER-SOLUBLE SILICON QUANTUM DOTS WITH QUASI-BLUE EMISSION

Yun Wang[1], Hao Wang[1], Jun Guo[2], Jiang Wu[3], Li J. Gao[1*], Ying H. Sun[1], J. Zhao[1] and Gui F. Zou[1*]

(1) College of Physics, Optoelectronics and Energy & Collaborative Innovation Center of Suzhou Nano Science and Technology, Soochow University

(2) Testing and Analysis Center, Soochow University

(3) State Key Laboratory of Electronic Thin Film and Integrated Devices, University of Electronic Science and Technology of China

ABSTRACT

In this study, water-soluble silicon quantum dots have quasi-blue emission at 390 nm by being capped with 1-vinylimidazole in resese micelles. As-obtained silicon quantum dots have a diameter of 2~5 nm and high crystallinity. The quasi-blue emission of the silicon quantum dots is likely attributed to the polarity of the capping ligands. Moreover, the silicon quantum dots are water-soluble and have photoluminescence nanosecond decay time, suggesting their potential application in biological field.

Keyword: Water-soluble Quasi-blue emission Silicon quantum dots

BACKGROUND

In the recent years, extensive research effects have been dedicated to silicon quantum dots (Si QDs) due to their potential applications in a wide range of areas, such as photovoltaic, [1–3] light-emitting devices [4, 5], and biological fields [6–8]. Si QDs show the strong quantum effects and the increasing photoluminescence (PL) intensity by radiative recombination and direct band gap transitions to reduce phonon-assisted indirect band gap transitions.

Over the last decades, researchers have made great efforts to obtain Si QDs with different emissive light. In general, their emission peak can be tuned by changing the particles size or their capping ligands [9–14]. The tunable emission wavelength of Si QDs in the visible range suggests that these materials are promising candidates for bio-applications. As reported, Si QDs with oxide surface passivation typically exhibit dipole-forbidden yellow-red emission [15, 16]. However, they have a long radiative lifetime (10^{-6}–10^{-3} s), which heavily limits their application in biological imaging. Fortunately, once passivated by a hydrogen or carbon atoms, the Si QDs will show blue emission and have electric-dipole-allowed direct band gap transitions which lead to a shorter radiative lifetime (10^{-9}–10^{-8} s) [15, 17] and are suitable for biological imaging. To date, many blue PL Si QDs were designed and synthesized by different methods [12, 18, 19]. However, only few of them were water-soluble [20], which still cannot nearly satisfy the growing need for biological study. Therefore, the requirement of blue PL stable and water-soluble Si QDs remains.

Herein, we report a kind of water-soluble and stable Si QDs by the selection of 1-vinylimidazole to coat silicon quantum dots in reverse micelles. The As-synthesized Si QDs manifest a quasi-blue light emission peak at 390 nm and red-shifted around 50 nm compared to 1-heptene-capped ones. Those obtained Si QDs were mono-dispersed with a size distribution below 5 nm and high crystallinity. All these characterizations indicated that the water-soluble Si QDs were feasible for further applications in biological areas.

METHODS

Synthesis and Purification of Alkyl-Si QDs

Si QDs were synthesized by the solution-phase reduction of $SiCl_4$ in reverse micelles as reported. All reagents and manipulations were carried out in an argon atmosphere in a glove box to prevent oxidation of Si QDs. Si QDs were formed by adding 5.2 ml of the reducing agent (dropwise), 1 M of lithium aluminum hydride ($LiAlH_4$) in tetrahydrofuran (THF) (strong reducing agent), to 50 ml of toluene containing 1 g of tetraoctylammonium bromide (TOAB) and 300 μl of $SiCl_4$ in a 100-ml round-bottomed flask (the solution has been stirred for 1 h). After finishing adding the reducing agent, the solution was left to react for 3 h. Then the solution was quenched with 50 ml of anhydrous methanol. After stirring for several minutes, 100 μl of 0.1 M hexachloroplatinic acid (H_2PtCl_6) in isopropyl alcohol was added in the mixture. The surface-capping process was performed by adding 0.7 ml of 1-vinylimidazole into the solution. The solution was left to react for another 3 h. After that, all the solvents were removed from the mixture by rotary evaporation. The resulting gray white powder was redispersed in 20 ml of water for further purification and/or solvent exchange. The solution was filtered through 0.45 and 0.22 μm membrane filter twice to remove the excessive TOAB, and then a clear solution of Si QDs in water was obtained. To 1-heptene-capped Si QDs, they were distributed in hexane and the received solution was also washed with 20 ml of n-methyl formamide and finally with distilled water.

Characterization

UV-vis absorption spectra were collected using a Shimadzu spectrophotometer (mode UV2450). PL and lifetime spectra were recorded by a Horiba spectrofluorometer (Fluoromax-4) using an excitation wavelength of 290 nm. Transmission electron microscope (TEM) studies were performed using a FEI Tecnai G2 F20 microscope operating at 200 kV. The particles size distribution was measured by Malvern Nano ZS.

RESULTS AND DISCUSSION

The two integrated reaction processes are presented as Scheme 1. As we know, 1-vinylimidazole is more water-soluble than 1-heptene, leading to 1-vinylimidazole-capped Si QDs hydrophilic and 1-heptene-capped Si QDs hydrophobic. As depicted in the TEM (Fig. 1), the As-synthesized-alkyl-capped Si QDs are almost spherical and the distribution of both 1-heptene- and 1-vinylimidazole-capped Si QDs are relatively mono-dispersed, showing the alkylation occurrence on the Si quantum dots surface to result in non-aggregation. From the high-resolution transmission electron microscope (HRTEM) images (Fig. 1 insets (upper)), a single particle possesses the clear lattice fringes (0.21 nm) of Si quantum dots. It is consistent with the (211) plane in diamond crystalline silicon. It suggests that the Si QDs with high quality are synthesized. From the dynamic light scattering (DLS) histogram of the As-prepared Si QDs (Fig. 1 insets (lower)), the size distribution of Si QDs manifests alkyl-capped silicon QDs in a solution having an average size of 3.2 ± 0.4 nm and 1.5 ± 0.3 nm, respectively, which are identical with the estimated results of TEM characterization.

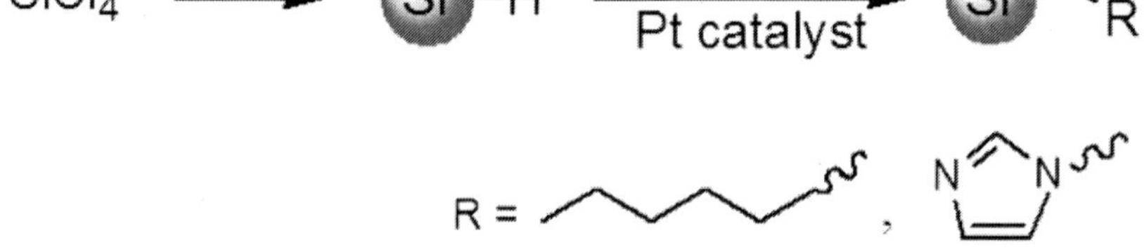

Scheme 1 Synthesis route of Si QDs capped with 1-heptene and 1-vinylimidazole, respectively

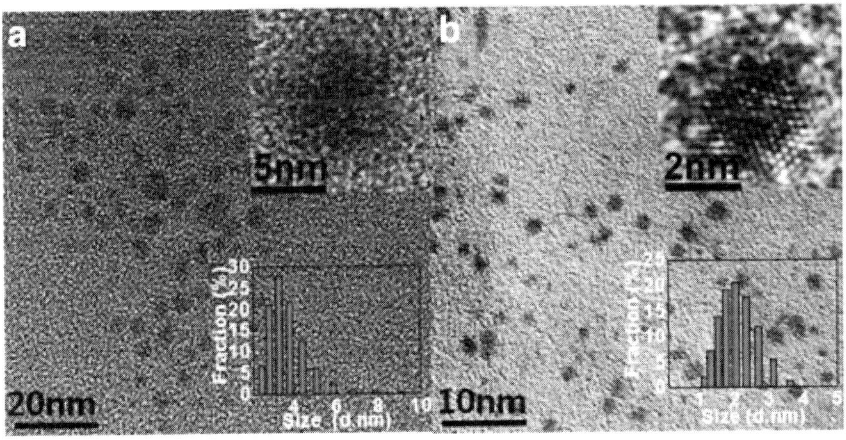

Figure. 1. TEM and HRTEM images of **A** 1-heptene-capped Si QDs and **B** 1-vinylimidazole-capped Si QDs. Insets (*upper*) show a single alkyl-capped Si QD with clear lattice fringes. Insets (*lower*) are size distribution histograms of 1-heptene-capped Si QDs and 1-vinylimidazole-capped Si QDs

It is well-known that light emission of Si nanostructures can be greatly affected by the surface ligand modification [21]. Even tiny differences in the styles or structures of ligands modification may lead to apparent changes in Si QDs PL. Moreover, there is a theoretical demonstration that 1–2 nm Si QDs with a hydrogen or carbon surface termination have direct band gap optical transitions. This will lead to PL staying in the UV/blue region of the electronic spectrum [9]. As shown in Fig. 2, an intense emission peak at 335 nm originates from the pristine silicon character. Nevertheless, when 1-heptene-capped Si QDs are excited by different wavelengths, two shoulder peaks at 321 nm and 350 nm may be caused by the inhomogeneous size distribution. When the Si surface is modified by 1-vinylimidazole, the Si QDs exhibit an intense and regular emission peak at 390 nm, which is red-shifted about 55 nm in comparison with the 1-heptene-capped ones. The difference of our Si QDs' emission peaks derives from the capping agents. As previously reported [22], the emission peak position is closely related with the capping ligands polarity. The emission peak of Si QDs capped with polar ligands is red-shifted compared with nonpolar capping ligands.

This is consistent with our results. To one's interest, there are no shoulder peaks along with the main Si emission peak. It can be indirectly told that the particles size seems to be more uniform, which is consistent with the above TEM characterizations.

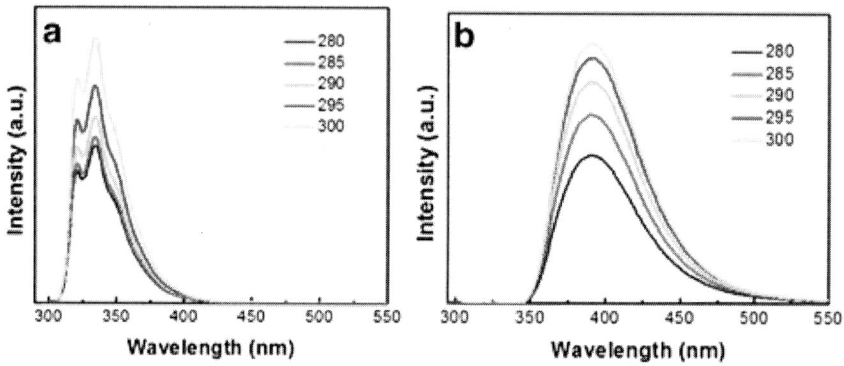

Figure. 2. PL spectra of **A** 1-heptene-capped Si QDs in hexane and **B** 1-vinylimidazole-capped Si QDs in water excited with different wavelengths from 280 to 300 nm with a 5-nm increment

As reported in previous literature, time-resolved spectroscopy reveals the lifetime of excited states before radiative recombination, and this provides insights into the recombination pathways in the material [23]. As shown in Fig. 3, the PL average decay of 1-heptene-capped Si QDs in hexane and 1-vinylimidazole-capped Si QDs in water are estimated to be 3.84 and 1.93 ns by two-exponential fit of both PL decay, respectively. The rapid rates of recombination measured also indicate that the observed emission results from dipole-allowed recombination across the direct band gap transition in silicon quantum dots with a carbon surface termination [20] Meanwhile, the PL lifetimes agree well with others' previous reports of PL decay measurements and with other reports of colloidal Si QDs with direct band gap emission [9, 17, 20].

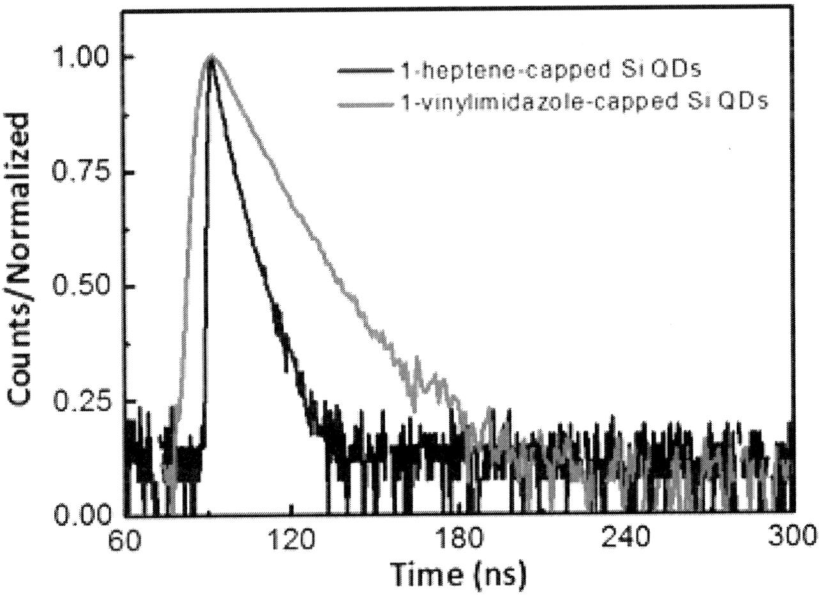

Figure. 3. Fluorescence decay as a function of time (ns) for 1-heptene-capped silicon QDs (*black line*) in hexane measured at an emission wavelength of 335 nm and 1-vinylimidazole-capped Si QDs (*red line*) at an emission wavelength of 390 nm. A laser source of 300 nm was used

As well known, the stability of Si QDs is very important to the following applications. In Fig. 4, PL intensity of Si QDs is studied at different times. In our experiments, PL spectra of 1-vinylimidazole-capped Si QDs are measured upon preparation, 1 day, and 2 days later, respectively. It can be seen that the PL intensity of 1-vinylimidazole-capped Si QDs decreased by 8.3 % after being synthesized 1 day later. While the PL intensity only decreases by 1.8 % after 2 days. It can tell that 1-vinylimidazole-capped Si QDs remain to be stable. Based on our experience, the alkenyl-imidazole or analog ligands may modify the Si QDs as well. This provides a platform to find the new phenomena for the modified Si QDs.

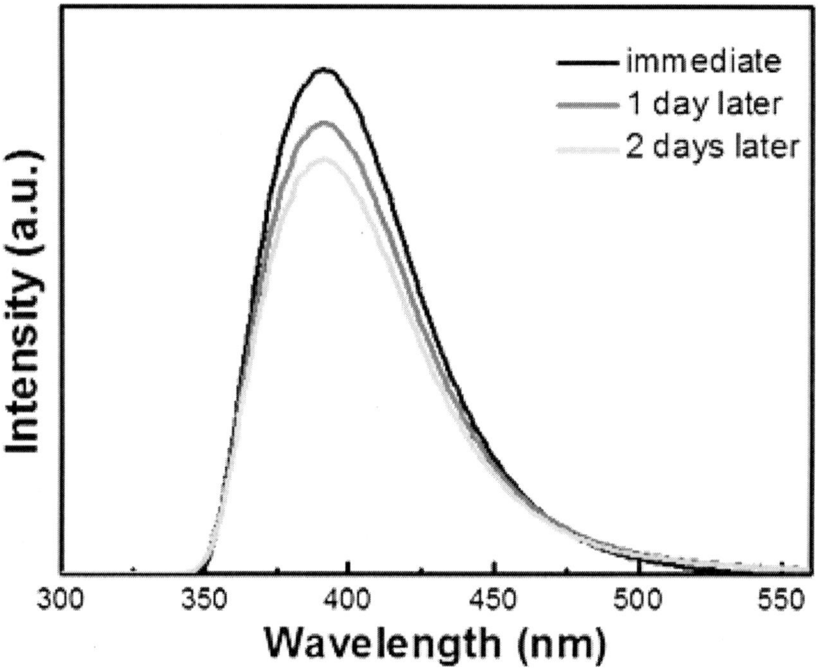

Figure. 4. PL stability of 1-vinylimidazole-capped Si QDs

Some researches about PL sensitivity to surrounding changes have been reported before [24, 25]. For example, as reported in a previous paper, the authors investigated nanosecond photoluminescence processes in colloidal core/shell CdSe/ZnS nanoparticles dissolved in water and found strong sensitivity of luminescence to the solvent state. Identically, the intensity of PL of our Si QDs are very strong, it is believable that they are sensitive to various changes in solvent, such as chemical and physical changes, via ligand reconstruction.

CONCLUSIONS

In summary, Si QDs with a size distribution within 2~5 nm are synthesized by a micro-emulsion colloidal solution. Using a capping ligand of 1-vinylimidazole, we can gain a water-soluble and quasi-blue light emissive (at 390 nm) Si QDs. Based on the test of time-resolved PL, it can be seen that the decay of 1-vinylimidazole-capped Si QDs are around 1.93 ns. The synthetic method may be extended to other similar capping ligands to Si QDs.

ABBREVIATIONS

DLS: dynamic light scattering

H_2PtCl_6 : hexachloroplatinic acid

HRTEM: high-resolution transmission electron microscope

$LiAlH_4$: lithium aluminum hydride

PL: photoluminescence

Si QDs: silicon quantum dots

TEM: transmission electron microscope

THF: tetrahydrofuran

TOAB: tetraoctylammonium bromide

ACKNOWLEDGEMENTS

We gratefully acknowledge the support from the "973 Program—the National Basic Research Program of China" Special Funds for the Chief Young Scientist (2015CB358600), the Excellent Young Scholar Fund from National Natural Science Foundation of China (21422103), the Jiangsu Fund for Distinguished Young Scientist (BK20140010), the Priority Academic Program Development of Jiangsu Higher Education Institutions (PAPD), and the Jiangsu Scientific and Technological Innovation Team (2013).

AUTHORS' CONTRIBUTIONS

HW and JW participated in the discussion of the study and helped to draft the manuscript. JZ and YS participated in the PL and UV test. JG carried out the TEM test. LG and GZ conceived of the study and participated in its design. All authors read and approved the final manuscript.

REFERENCES

1. Conibeer G, Green M, Cho EC, König D, Cho YH, Fangsuwannarak T, et al. Silicon quantum dot nanostructures for tandem photovoltaic cells. Thin Solid Films. 2008;516:6748–56.
2. Conibeer G, Green MA, König D, Perez-Wurfl I, Huang S, Hao X, et al. Silicon quantum dot based solar cells: addressing the issues of doping, voltage and current transport. Prog Photovolt: Res Appl. 2011;19:813–24.
3. Cho EC, Park S, Hao X, Song D, Conibeer G, Park SC, et al. Silicon quantum dot/crystalline silicon solar cells. Nanotechnology. 2008;19:245201–5.
4. Ghosh B, Masuda Y, Wakayama Y, Imanaka Y, Inoue J, Hashi K, et al. Hybrid white light emitting diode based on silicon nanocrystals. Adv Funct Mater. 2014;24:7151–60.
5. Pillai S, Catchpole KR, Trupke T, Zhang G, Zhao J, Green MA. Enhanced emission from Si-based light-emitting diodes using surface plasmons. Appl Phys Lett. 2006;88:161102–5.
6. Li ZF, Ruckenstein E. Water-soluble poly(acrylic acid) grafted luminescent silicon nanoparticles and their use as fluorescent biological staining labels. Nano Lett. 2004;4:1463–7.

7. Erogbogbo F, Yong KT, Roy I, Xu G, Prasad PN, Swihart MT. Biocompatible luminescent silicon quantum dots for imaging of cancer cells. ACS Nano. 2008;2:873–8.

8. Fujioka K, Hiruoka M, Sato K, Manabe N, Miyasaka R, Hanada S, et al. Luminescent passive-oxidized silicon quantum dots as biological staining labels and their cytotoxicity effects at high concentration. Nanotechnology. 2008;19:415102–7.

9. English DS, Pell LE, Yu Z, Barbara PF, Korgel BA. Size tunable visible luminescence from individual organic monolayer stabilized silicon nanocrystal quantum dots. Nano Lett. 2002;2:681–5.

10. Shiohara A, Prabakar S, Faramus A, Hsu CY, Lai PS, Northcote PT, et al. Sized controlled synthesis, purification, and cell studies with silicon quantum dots. Nanoscale. 2011;3:3364–70.

11. Kang Z, Liu Y, Tsang CHA, Ma DDD, Fan X, Wong NB, et al. Water-soluble silicon quantum dots with wavelength-tunable photoluminescence. Adv Mater. 2009;21:661–4.

12. Holmes JD, Ziegler KJ, Doty RC, Pell LE, Johnston KP, Korgel BA. Highly luminescent silicon nanocrystals with discrete optical transitions. J Am Chem Soc. 2001;123:3743–8.

13. Zou J, Baldwin RK, Pettigrew KA, Kauzlarich SM. Solution synthesis of ultrastable luminescent siloxane-coated silicon nanoparticles. Nano Lett. 2004;4:1181–6.

14. Warner JH, Rubinsztein-Dunlop H, Tilley RD. Surface morphology dependent photoluminescence from colloidal silicon nanocrystals. J Phys Chem B. 2005;109:19064–7.

15. Zhou Z, Brus L, Friesner R. Electronic structure and luminescence of 1.1- and 1.4-nm silicon nanocrystals: oxide shell versus hydrogen passivation. Nano Lett. 2003;3:163–7.

16. Puzder A, Williamson AJ, Grossman JC, Galli G. Computational studies of the optical emission of silicon nanocrystals. J Am Chem Soc. 2003;125:2786–91.

17. Wilcoxon JP, Samara GA, Provencio PN. Optical and electronic properties of Si nanoclusters synthesized in inverse micelles. Phys Rev B. 1999;60:2704–14

18. Švrček V, Mariotti D, Kondo M. Ambient-stable blue luminescent silicon nanocrystals prepared by nanosecond-pulsed laser ablation in water. Opt Express. 2009;17:521–8.

19. Hua F, Erogbogbo F, Swihart MT, Ruckenstein E. Organically capped silicon nanoparticles with blue photoluminescence prepared by hydrosilylation followed by oxidation. Langmuir. 2006;22:4363–70

20. Warner JH, Hoshino A, Yamamoto K, Tilley RD. Water-soluble photoluminescent silicon quantum dots. Angew Chem Int Ed. 2005;44:4550–4.

21. Kanemitsu Y, Futagi T, Matsumoto T, Mimura H. Origin of the blue and red photoluminescence from oxidized porous silicon. Phys Rev B. 1994;49:14732–5.

22. Shiohara A, Hanada S, Prabakar S, Fujioka K, Lim TH, Yamamoto K, et al. Chemical reactions on surface molecules attached to silicon quantum dots. J Am Chem Soc. 2010;132:248–53.

23. Warner JH, Tilley RD. Synthesis of water-soluble photoluminescent germanium nanocrystals. Nanotechnology. 2006;17:3745–9.

24. Antipov A, Bell M, Yasar M, Mitin V, Scharmach W, Swihart M, et al. Luminescence of colloidal CdSe/ZnS nanoparticles: high sensitivity to solvent phase transitions. Nanoscale Res Lett. 2011;6:142–8

25. Pihlasalo S, Kirjavainen J, Hänninen P, Härmä H. High sensitivity luminescence nanoparticle assay for the detection of protein aggregation. Anal Chem. 2011;83:1163–6.

3

CHAPTER

NITROGEN-DOPED CARBON DOTS FOR "GREEN" QUANTUM DOT SOLAR CELLS

Hao Wang[1], Pengfei Sun[1], Shan Cong[1], Jiang Wu[2], Lijun Gao[1]*, Yun Wang[1], Xiao Dai[1], Qinghua Yi[1] and Guifu Zou[1]*

(1)College of Physics, Optoelectronics and Energy & Collaborative Innovation Center of Suzhou Nano Science and Technology, Soochow University

(2)Department of Electronic and Electrical Engineering, University College London

ABSTRACT

Considering the environment protection, "green" materials are increasingly explored for photovoltaics. Here, we developed a kind of quantum dots solar cell based on nitrogen-doped carbon dots. The nitrogen-doped carbon dots were prepared by direct pyrolysis of citric acid and ammonia. The nitrogen-doped carbon dots' excitonic absorption depends on the N-doping content in the carbon dots. The N-doping can be readily modified by the mass ratio of reactants. The constructed "green" nitrogen-doped carbon dots solar cell achieves the best power conversion efficiency of 0.79 % under AM 1.5 G one full sun illumination, which is the highest efficiency for carbon dot-based solar cells.

Keywords: Carbon dots Nitrogen-doped Quantum dot solar cells

BACKGROUND

Among the third-generation photovoltaics, quantum dot solar cells (QDSCs) are emerging as a promising candidate due to the unique and versatile characteristics of quantum dots (QDs) including tunable band gap and high absorption coefficient [1–4]. Typically, low-band gap metal chalcogenide (CdS, CdSe, CdTe, PbS, PbSe, $CuInS_2$, etc.) QDs are widely used as sensitizers in QDSCs [5–10]. The QDSCs have been well developed, but they contain highly toxic metals (including Cd, Pb, and In). Hence, the environmentally friendly alternatives ("green" materials) are extremely needed and welcomed to the fabrication of solar cells.

Due to the unique optical and electronic properties, water solubility, excellent biocompatibility, low toxicity, and robust chemical inertness [11–15], carbon dots (CDs) as an emerging carbon-based nanomaterial have been widely researched and applied in many fields [16–22]. One of the CDs applications is tentatively used as sensitizers for QDSCs [23–26]. For instance, Mirtchev et al. [24] have prepared water soluble CDs as sensitizers for QDSCs achieving power conversion efficiency (PCE) of 0.13 %. Zhang et al. [25] have also fabricated 0.13 % PCE of QDSCs based on nitrogen-doped CDs (NCDs). Up to now, the low short-circuit current density (J_{sc}) might be the main factor limiting the efficiency of the cells in comparison with the open-circuit voltage (V_{oc}) and fill factor (FF) of metal chalcogenide-based QDSCs. Narrow light-absorption and diverse trap states result in the extremely low J_{sc} of CD-based QDSCs. Meanwhile, CDs' light-absorption is centered at ultra-violet region and very weak at visible region. Because the existence of trap state defects with different energy levels has been demonstrated by their excitation-dependent fluorescence [27], partially photoinduced carriers will be recombined by CDs rather than being injected into the transporting media. Accordingly, the PCE of QDSCs is very low. Much effort still needs to improve the practicable sensitizers for QDSCs.

In this study, we have developed NCDs to build up the "green" QDSCs with enhanced efficiency. The NCDs were prepared by direct pyrolysis of citric acid and ammonia. Experimental results show that the NCDs' excitonic absorption depends on the N-doping content modified by the mass ratio of reactants. The NCDs with optimal excitonic absorption were used as sensitizers in the QDSCs. The constructed NCD solar cell achieves the best PCE of 0.79 % under AM 1.5 G one full sun illumination. It is noting that the obtained efficiency is the highest one in the reported solar cell-based CDs.

METHODS

Synthesis

The NCDs were prepared by a direct pyrolysis method. As shown in Scheme 1, the NCDs are prepared by direct pyrolysis of citric acid (CA, Alfa Aesar) and ammonia (Sinopharm Chemical Reagent Co.) mixture similar to the previous report [28]. Typically, CA (4 g) and certain amount of ammonia are dissolved in distilled water (10 mL) with stirring for 30 min. Then, the prepared solution is transferred into a porcelain boat and then heated in air at 200 °C for 3 h with a heating rate of 10 °C/min. The obtained carbogenic product is mixed with acetone (40 mL, Sinopharm Chemical Reagent Co.) by ultrasonic. The supernatants containing NCDs are collected after centrifugation at 10,000 rpm for 15 min. Then, the acetone is removed by rotary evaporation, and solid-state NCDs are collected for further characterization and usage.

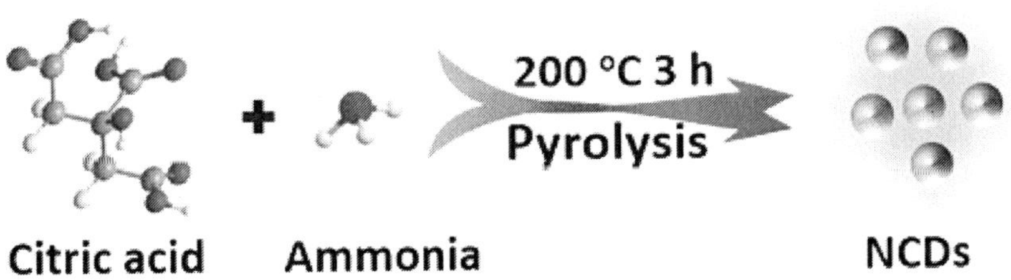

Scheme 1 Preparation of NCDs via direct pyrolysis of citric acid and ammonia

Fabrication of Solar Cell

The NCD-sensitized TiO_2 photoanodes are fabricated according to the previous report [29]. The cleaned fluorine-doped tin oxide (FTO) glasses are immersed in 40 mM $TiCl_4$ aqueous solution at 70 °C for 30 min and washed with water and ethanol. A 20-nm-sized TiO_2 paste is deposited on the FTO glass plate by doctor blade printing technique and then dried at 125 °C for 5 min. The scattering layer of 200-nm-sized TiO_2 paste is coated on the top of the first TiO_2 layer, followed by sintering in air at 500 °C for 30 min. After cooling to 95 °C, the TiO_2 film is immersed in the NCQD acetone solution at room temperature for 2 h. The NCD-sensitized TiO_2 is washed with anhydrous ethanol and dried with nitrogen stream. The solar cells are fabricated by sandwiching gel electrolytes between a NCD-sensitized TiO_2 electrode and a Pt counter electrode, which are separated by a 25-mm-thick hot-melt ring (Surlyn, Dupont) and sealed by heating. The electrolyte injection hole on the thermally platonized FTO counter electrode is finally sealed with Surlyn sheet and a thin glass by heating.

Characterization and Measurement

The morphology of the materials is investigated by transmission electron microscopy (TEM, FEI Tecnai G-20) and atomic force microscopy (AFM, Asylum Research MFP-3D-BIO). Fourier transform infrared (FT-IR) spectra are obtained on a Bruker VERTEX70 FT-IR spectrometer ranged from 4000 to 400 cm^{-1}. X-ray photoelectron spectra (XPS) are acquired with a Japan Kratos Axis Ultra HAS spectrometer using a monochromatic Al Kα source. The crystallinity of NCDs were characterized by Raman spectroscopy (HORIBA Jobin Yvon HR800) using 514-nm laser as the excitation source. The UV–visible absorption spectrum and diffuse-reflectance spectra are recorded on a UV2501PC (Shimadzu). The photoluminescence (PL) spectra are recorded on a Fluoromax-4 spectrofluorometer (HORIBA JobinYvon Inc.) equipped with a 150 W of xenon lamp as the excitation source. Photocurrent–voltage measurements of solar cells (Keithley2440 sourcemeter) are obtained by using a solar simulator (Newport) with an AM 1.5 G filter under an irradiation intensity of 100 mWcm^{-2} (550-W Xe source, Abet). The light intensity is calibrated using a standard silicon photovoltaic solar cell. The active cell area is 0.25 cm^2. All of the measurements above are performed at room temperature.

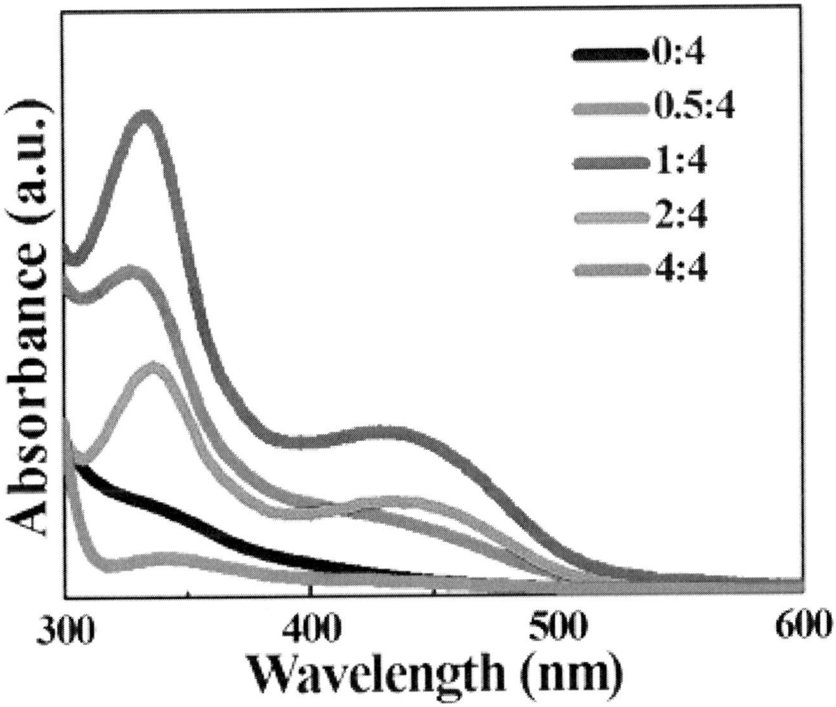

Figure. 1. UV–vis spectra of NCDs from 4 g CA and different amounts (0, 0.5, 1, 2, and 4 g) of ammonia

RESULTS AND DISCUSSION

The UV–vis spectra of NCDs from different reactant mass ratios show the NCDs' excitonic absorption strongly depends on their N-doping content (Fig. 1). Our results demonstrate that the N-doping content is easily modified by the mass ratio of reactants. When the mass ratio of the reactants ammonia and CA is around 1: 4, the NCDs possess the optimal visible absorption. The excessive ammonia might happen strong chemical reaction with overheated NCDs resulting in excessive amount of water loss and the destruction of NCDs surface. To explore the effect of the mass ratio on the N-doping level of the NCDs, the element constitutes of the NCDs have been analyzed in Table 1. The N-doping content increases with the increasing mass ratio of ammonia and CA. Nevertheless, it changes slight when the mass ratio further increases over 4:1.

Table 1. Relative elemental analysis of the CDs with different N-doping levels by XPS analysis

Elemental/atomic (%)	C	O	N
N-free	78.4	21.6	0
0:4	74.8	22.6	2.6
1:4	67.5	22.5	10.0
2:4	69.7	21.1	9.7
4:4	71.5	19.2	9.3

The optical properties of optimal NCDs were further studied in Fig. 2. Compared with N-free CDs, the optimal NCDs exhibit not only much higher absorption peak around 335 nm but also more broad absorption band extending to 550 nm (Fig. 2a). In general, the absorption peak around 335 nm corresponds to n–π* transition of C=O bonds while the visible absorption band is attributed to the amino groups on the surface of NCDs [30]. The expanding absorption to 550 nm indirectly explains nitrogen-doping CDs. In addition, NCDs also exhibit excellent solubility with bright blue emission under UV lamp (365 nm) irradiation (inset of Fig. 2a). With the excitation wavelength increasing from 300 to 500 nm in Fig. 2b, the emission intensity gradually decreases and the emission peak shifts from 425 to 510 nm [31]. The surface groups of NCDs introduce trap states with different energy levels resulting in the excitation-dependent emission. The NCDs' fluorescence quantum yield (FLQY) was calculated to be ~36 % with quinine sulfate (FLQY 54 %) as the reference (see Additional file 1: Figure S1).

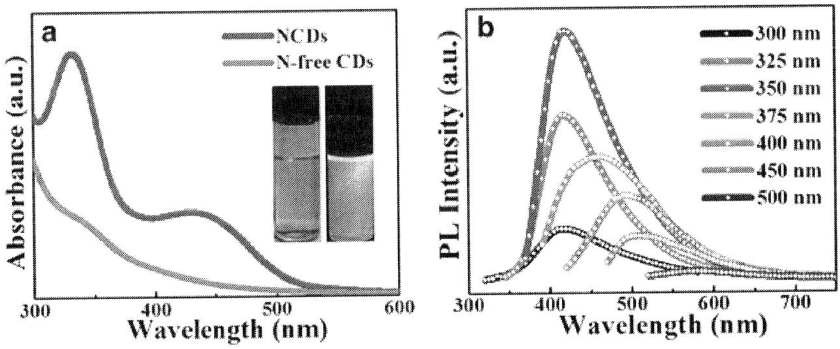

Figure. 2. **A** UV–vis spectra of NCDs and N-free CDs. *Insets* are the photographs of NCDs aqueous solution under day light (*left*) and UV lamp irradiation (*right*). **B** PL spectrum of NCDs with excitation of different wavelengths recorded from 300 to 500 nm

NCDs were characterized by FT-IR spectroscopy (see Additional file 1: Figure S2). The absorption bands of C–N stretching vibrations at 1182 cm^{-1} confirms the doping of N in the NCDs. Besides, there also observe O–H stretching vibrations at 3000–3400 cm^{-1}, C–H stretching vibrations at 2940 cm^{-1}, C=O stretching vibrations at 1710 cm^{-1}, C–O stretching vibrations at 1408 and 1286 cm^{-1}, C–O–C stretching vibrations at 1350 cm^{-1}. XPS measurements were performed for the surface elemental analysis of NCDs in Fig. 3a. The full-scan spectrum reveals the existence of carbon (C 1 s, 284.5 eV), nitrogen (N 1 s, 400.5 eV), and oxygen (O 1 s, 531.5 eV). In the expanded XPS spectrum (Fig. 3b), carbon's 1-s peaks at 284.6, 285.7, and 288.1 eV can be assigned to bonds of C–C, C–N/C–O, and C=O, respectively. Figure 3c illustrates nitrogen 1-s peaks at 400.3 and 401.6 eV suggesting nitrogen exists in the form of C–N and N–H bonds. As shown in Fig. 3d, oxygen 1-s peak exhibits two peaks at 531.7 and 532.8 eV, which are attributed to C=O and C–OH/C–O–C, respectively. It can be concluded that nitrogen-doped CDs have a variety of polar groups including hydroxyl, alkyl, and carboxyl [32].

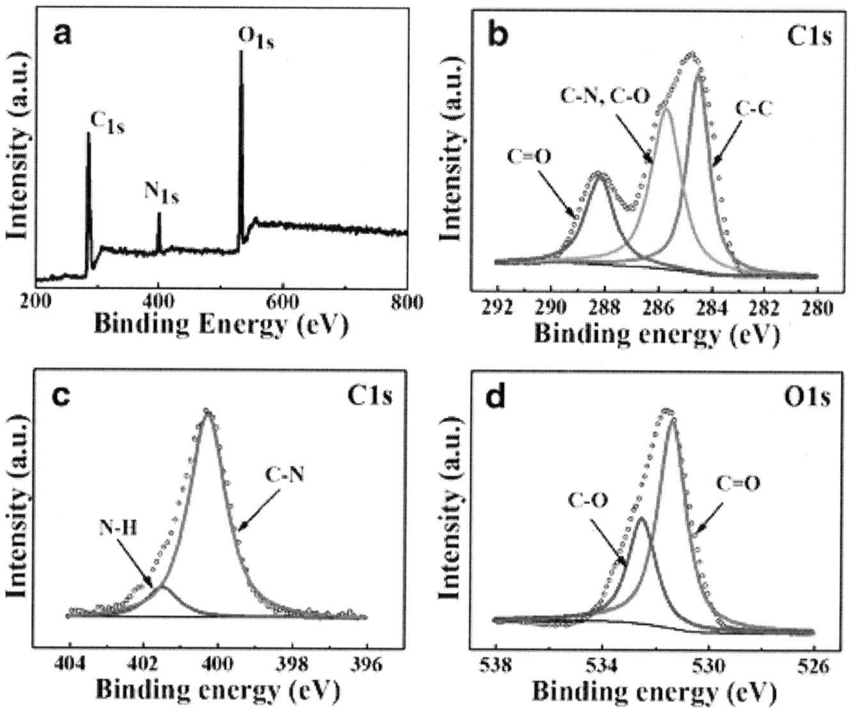

Figure. 3. XPS survey scan of NCDs on indium foil. **A** XPS full scan spectrum. XPS high resolution survey scan of (**B**) C1s, (**C**) N1s, and (**D**) O1s region

The morphology of NCDs was characterized by TEM and AFM in Fig. 4. The TEM images (Fig. 4a, b) show that NCDs' sizes are mainly distributed in the range of 7–15 nm with an average size of 10.8 nm. The AFM image (Fig. 4c–e) depicts that the topographic height of the NCDs is in the range of 1–3 nm with main distribution at 2 nm. It suggests the NCDs have the cylindrical-like structure (similar to small pieces of graphitic layer stacking) [32]. Raman spectroscopy was used to analyze the relative intensity ratio about 1.07 of the D-band (1342 cm^{-1}) and G-band (1531 cm^{-1}) (see Additional file 1: Figure S3).

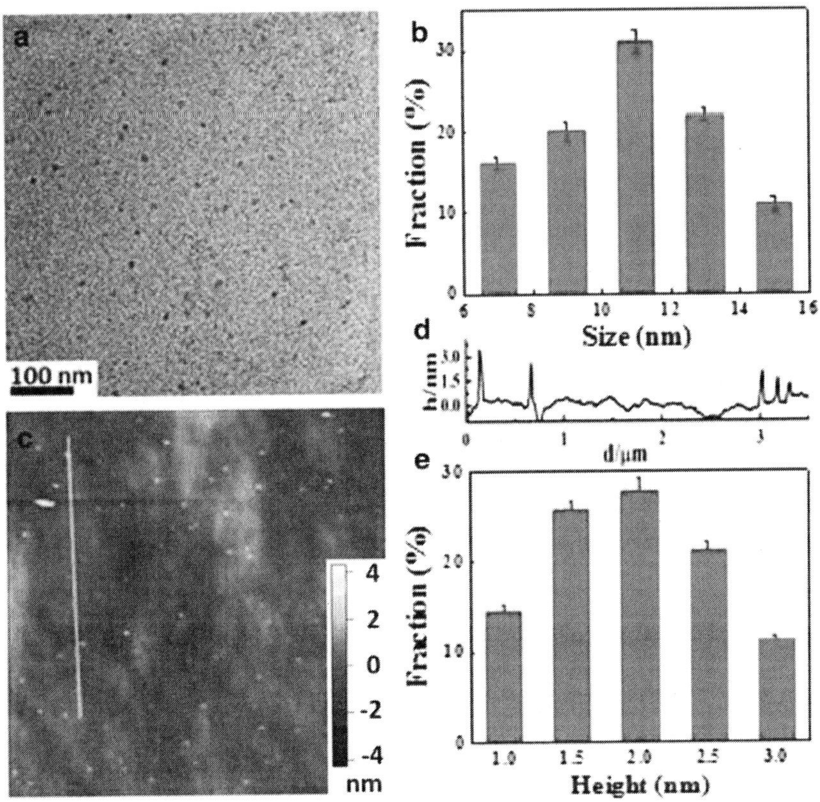

Figure. 4. A TEM image of NCDs. **B** Particle size distribution of NCDs. **C** AFM image of NCDs deposited on silicon substrate. **D** Height profile along the white line in (**C**). **E** Height distribution of NCDs

The enhanced visible-light absorption delivers a message that NCDs could be an attractive interface material integrated with porous TiO_2 for QDSCs. As a demonstration, NCQDs were combined with anatase TiO_2. Figure 5a describes UV–vis diffuse reflectance spectrum of TiO_2 (inset: NCDs/TiO_2 photoanode). It could be seen that NCD/TiO_2

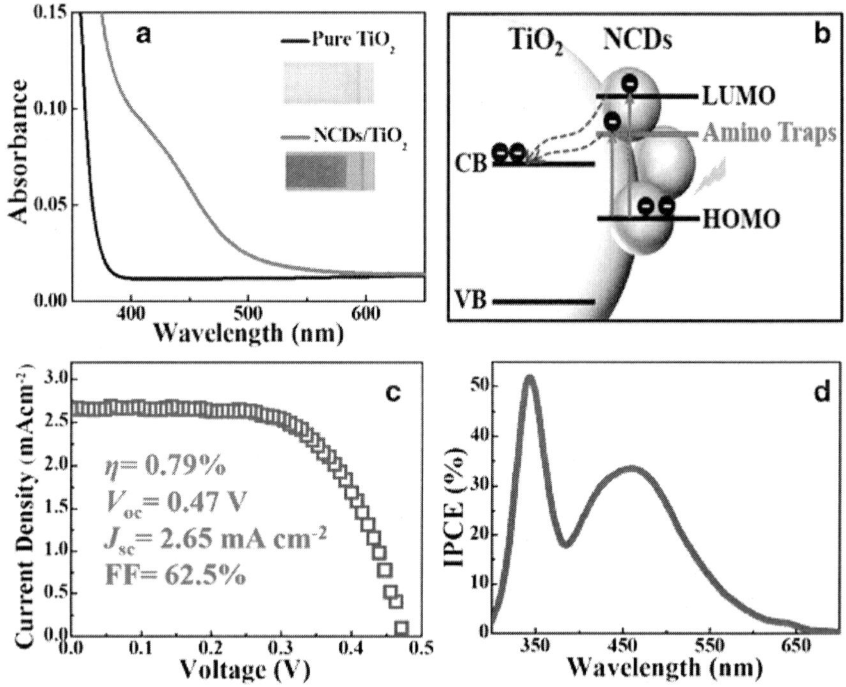

Figure. 5. **A** UV–vis DRS spectra of TiO_2 and NCD/TiO_2 photoanode. *Insets* are the photographs of TiO_2 and NCD/TiO_2 photoanode. **B** A proposed model of photoinduced electron transition between NCDs and TiO_2. **C** Current density–voltage characteristics and (**D**) IPCE action spectrum of NCD-sensitized solar cells

photoanode gains much more enhanced absorption in visible region ranging from 400 to550 nm than the pure TiO_2 photoanode. It should be helpful to apply for photosensitizers. Furthermore, the PL spectrum of NCDs, TiO_2, and NCDs/TiO_2 were also studied to confirm the phenomena (see Additional file 1: Figure S4). Under visible light excitation of 420 nm, pure TiO_2 has negligible PL emission while NCDs exhibit a strong and wide emission peak centered at 550 nm. However, it is interesting to see that the emission of NCDs/TiO_2 greatly weakened. It indicates that TiO_2 impedes the recombination of photogenerated electrons and holes in excited NCDs and absorbs the photogenerated electrons. The schematic diagram in Fig. 5b presents the photoinduced electrons transition between NCDs and TiO_2. Compared to N-free CDs, the N-dopant of NCDs can introduce an additional energy level between π of carbon and π* of oxygen. Once excited, NCDs can absorb the relative low-energy photons at visible light range and generate the photoexcited electrons. As a result, there are more electrons transferring to the conducting band of the TiO_2, resulting in the higher current density. To investigate the potential application in photovoltaics, we fabricated NCDs as sensitizers for QDSCs. Figure 5c shows the current density–voltage (J-V) curve of NCDs sensitized solar cells under AM 1.5 G one full sun illumination. The solar cell has a J_{sc} of 2.65 mA cm^{-2}, a V_{oc} of 0.47 V, a FF of 62.5 %, and overall PCE of 0.79 %. To our best knowledge, the PCE is the highest value in the reported QDSCs based on CDs. Incident photon-to-electron conversion efficiency (IPCE) is plotted in Fig. 5d. It was found that the overall photocurrent response closely matches the corresponding absorption spectrum of NCD-sensitized TiO_2 film in Fig. 5a. IPCE value is over 10 % between 400 and 550 nm. It is noting that the highest value of IPCE even approaches 34 % for NCDs.

CONCLUSIONS

In summary, we have developed "green" quantum dot solar cells based on nitrogen-doped carbon dots. The optimal NCD-based solar cells achieve a PCE of 0.79 % with J_{sc} of 2.65 mA cm^{-2}, V_{oc} of 0.47 V, and FF of 62.5 %. The obtained PCE is the highest value in the reported QDSCs based on CDs. The study demonstrates that the further optimization of nitrogen dopant can gain a better conversion efficiency of NCD-based solar cells.

ACKNOWLEDGEMENTS

We gratefully acknowledge the support from "973 Program – the National Basic Research Program of China" Special Funds for the

Chief Young Scientist (2015CB358600), the Excellent Young Scholar Fund from National Natural Science Foundation of China (21422103), Jiangsu Fund for Distinguished Young Scientist (BK20140010), Jiangsu Specially-Appointed Professor Program (SR10800412), the Priority Academic Program Development of Jiangsu Higher Education Institutions (PAPD), and Jiangsu Scientist and Technological Innovation Team (2013).

AUTHORS' CONTRIBUTIONS

HW carried out all the measurement and drafted the manuscript. PS and SC fabricated the solar cells and analyzed the performance. JW and XD participated in the discussion of the study and helped to write the manuscript. YW carried out the TEM test and QY carried out the AFM test. LG and GZ conceived the study and participated in its design. All authors read and approved the final manuscript.

REFERENCES

1. Semonin E, Luther JM, Choi S, Chen HY, Gao J, Nozik AJ, Beard MC (2011) Peak external photocurrent quantum efficiency exceeding 100 % via MEG in a quantum dot solar cell. Science 334:1530–1533
2. Sargent EH (2012) Colloidal quantum dot solar cells. Nat Photonics 6:133–135
3. Kramer IJ, Sargent EH (2013) The architecture of colloidal quantum dot solar cells: materials to devices. Chem Rev 114:863–882
4. Kim MR, Ma D (2014) Quantum-dot-based solar cells: recent advances, strategies, and challenges. J Phys Chem Lett 6:85–99
5. Santra PK, Kamat PV (2012) Tandem-layered quantum dot solar cells: tuning the photovoltaic response with

luminescent ternary cadmium chalcogenides. J Am Chem Soc 135:877–885

6. Zhu H, Song N, Lian T (2013) Charging of quantum dots by sulfide redox electrolytes reduces electron injection efficiency in quantum dot sensitized solar cells. J Am Chem Soc 135:11461–11464

7. Jiao S, Shen Q, Mora-Seró I, Wang J, Pan Z, Zhao K, Kuga Y, Zhong X, Bisquert J (2015) Band engineering in core/shell ZnTe/CdSe for photovoltage and efficiency enhancement in exciplex quantum dot sensitized solar cells. ACS Nano 9:908–915

8. Zhang J, Gao J, Church CP, Miller EM, Luther JM, Klimov VI, Beard MC (2014) PbSe quantum dot solar cells with more than 6 % efficiency fabricated in ambient atmosphere. Nano Lett 14:6010–6015

9. Speirs MJ, Balazs DM, Fang HH, Lai LH, Protesescu L, Kovalenko MV, Loi MA (2015) Origin of the increased open circuit voltage in PbS-CdS core-shell quantum dot solar cells. J Mater Chem A 3:1450–1457

10. Pan Z, Mora-Seró I, Shen Q, Zhang H, Li Y, Zhao K, Wang J, Zhong X, Bisquert J (2014) High-efficiency "green" quantum dot solar cells. J Am Chem Soc 136:9203–9210

11. Lim SY, Shen W, Gao Z (2015) Carbon quantum dots and their applications. Chem Soc Rev 44:362–381

12. Ge J, Lan M, Zhou B, Liu W, Guo L, Wang H, Jia Q, Niu G, Huang X, Zhou H, Meng X, Wang P, Lee CS, Zhang W, Han X (2014) A graphene quantum dot photodynamic therapy agent with high singlet oxygen generation. Nat Commun 5:4596

13. Baker SN, Baker GA (2010) Luminescent carbon nanodots: emergent nanolights. Angew Chem Int Ed 49:6726–6744

14. Liu RL, Wu DQ, Liu SH, Koynov K, Knoll W, Li Q (2009) An aqueous route to multicolor photoluminescent carbon dots using silica spheres as carriers. Angew Chem Int Ed 48:4598–4601

15. Zhao QL, Zhang ZL, Huang BH, Peng J, Zhang M, Pang DW (2008) Facile preparation of low cytotoxicity fluorescent carbon nanocrystals by electrooxidation of graphite. Chem Commun 41:5116–5118.

16. Wang L, Wang Y, Xu T, Liao H, Yao C, Liu Y, Li Z, Chen Z, Pan D, Sun L, Wu M (2014) Gram-scale synthesis of single-crystalline graphene quantum dots with superior optical properties. Nat Commun

17. Ding C, Zhu A, Tian Y (2014) Functional surface engineering of C-dots for fluorescent biosensing and in vivo bioimaging. Acc Chem Res 47:20–30

18. Zhu Z, Ma J, Wang Z, Mu C, Fan Z, Du L, Bai Y, Fan L, Yan H, Phillips DL (2014) Efficient enhancement of perovskite solar cells through fast electron extraction: the role of graphene quantum dots. J Am Chem Soc 136:3760–3763

19. Wang H, Gao P, Wang Y, Guo J, Zhang KQ, Du D, Dai X, Zou G (2015) Fluorescently tuned nitrogen-doped carbon dots from carbon source with different content of carboxyl groups. APL Mater 3:086102

20. Wang H, Wang Y, Guo J, Su Y, Sun C, Zhao J, Luo H, Dai X, Zou G (2015) A new chemosensor for Ga^{3+} detection by fluorescent nitrogen-doped graphitic carbon dots. RSC Adv 5:13036–13041

21. Li H, He X, Kang Z, Huang H, Liu Y, Liu J, Lian S, Tsang CHA, Yang X, Lee ST (2010) Water-soluble fluorescent carbon quantum dots and photocatalyst design. Angew Chem Int Ed 49:4430–4434

22. Mao LH, Tang WQ, Deng ZY, Liu SS, Wang CF, Chen S (2014) Facile access to white fluorescent carbon dots toward light-emitting devices. Ind Eng Chem Res 53:6417–6425

23. Yan X, Cui X, Li B, Li LS (2010) Large, solution-processable graphene quantum dots as light absorbers for photovoltaics. Nano Lett 10:1869–1873

24. Mirtchev P, Henderson EJ, Soheilnia N, Yip CM, Ozin GA (2012) Solution phase synthesis of carbon quantum dots as

sensitizers for nanocrystalline TiO_2 solar cells. J Mater Chem 22:1265–1269

25. Zhang YQ, Ma DK, Zhang YG, Chen W, Huang SM (2013) N-doped carbon quantum dots for TiO_2-based photocatalysts and dye-sensitized solar cells. Nano Energy 2:545–552

26. Sun M, Ma X, Chen X, Sun Y, Cui X, Lin Y (2014) A nanocomposite of carbon quantum dots and TiO_2 nanotube arrays: enhancing photoelectrochemical and photocatalytic properties. RSC Adv 4:1120–1127

27. Li X, Zhang S, Kulinich SA, Liu Y, Zeng H (2014) Engineering surface states of carbon dots to achieve controllable luminescence for solid-luminescnet composites and sensitive Be^{2+} detection. Sci Rep 4:4976

28. Dong Y, Shao J, Chen C, Li H, Wang R, Chi Y, Lin X, Chen G (2012) Blue luminescent graphene quantum dots and graphene qxide prepared by tuning the carbonization degree of citric acid. Carbon 50:4738–474

29. Li Q, Zhao J, Sun B, Lin B, Qiu L, Zhang Y, Chen X, Lu J, Yan F (2012) High-temperature solid-state dye-sensitized solar cells based on organic ionic plastic crystal electrolytes. Adv Mater 24:945–950

30. Zhang X, Zhang Y, Wang Y, Kalytchuk S, Kershaw SV, Wang Y, Wang P, Zhang T, Zhao Y, Zhang H, Cui T, Wang Y, Zhao J, Yu WW, Rogach AL (2013) Color-switchable electroluminescence of carbon dot light-emitting diodes. ACS Nano 7:11234–11241

31. Wu ZL, Zhang P, Gao MX, Liu CF, Wang W, Leng F, Huang CZ (2013) One-pot hydrothermal synthesis of highly luminescent nitrogen-doped amphoteric carbon dots for bioimaging from Bombyx mori silk-natural proteins. J Mater Chem B 1:2868–2873

32. Li Y, Zhao Y, Cheng H, Hu Y, Shi G, Dai L, Qu L (2011) Nitrogen-doped graphene quantum dots with oxygen-rich functional groups. J Am Chem Soc 134:15–18

4

CHAPTER

THEORETICAL AND EXPERIMENTAL STUDIES OF (IN,GA)AS/GAP QUANTUM DOTS

Cedric Robert[1*], Tra Nguyen Thanh[1], Charles Cornet[1], Pascal Turban[2], Mathieu Perrin[1], Andrea Balocchi[3],

(1)Université Européenne de Bretagne, INSA Rennes, France CNRS, UMR 6082 Foton-Ohm

(2)Equipe de Physique des Surfaces et Interfaces, Institut de Physique de Rennes UMR UR1-CNRS 6251, Université de Rennes 1

(3)Université de Toulouse, INSA-CNRS-UPS, LPCNO

(4)Ioffe Physico-Technical Institute, Russian Academy of Sciences

ABSTRACT

(In,Ga)As/GaP(001) quantum dots (QDs) are grown by molecular beam epitaxy and studied both theoretically and experimentally. The electronic band structure is simulated using a combination of **K·P** and tight-binding models. These calculations predict an indirect to direct crossover with the In content and the size of the QDs. The optical properties are then studied in a low-In-content range through photoluminescence and time-resolved photoluminescence experiments. It suggests the proximity of two optical transitions of indirect and direct types.

Keywords: Quantum dots Tight-binding **K·P** simulation Time-resolved photoluminescence 78.55.Cr 78.47.jd 78.67.Hc

BACKGROUND

In the context of the monolithic integration of photonics on silicon, the pseudomorphic approach, i.e., growing lattice-matched compounds on Si, is a promising route towards an efficient and long-term stable laser on Si[1]. It should overcome the issue of the dramatic number of crystalline defects due to the large lattice mismatch encountered in the growth of most III-V materials onto Si substrates[2]. Among binary III-V materials, GaP presents the closest lattice constant to Si (0.37% at 300 K). The perfect lattice matching can even be obtained by introducing 2% of nitrogen in GaP. Recently, the epitaxial growth on Si substrate of GaP and $GaPN_{0.02}$ has been greatly improved by several groups[3–5]. Various active zones grown on GaP substrate or on $GaPN_{0.02}$/GaP/Si have been proposed. The best results have been achieved with compressive strained GaNAsP/GaP quantum wells (QWs) in electrically pumped lasers operating up to 150 K (Si substrate)[6] or at room temperature (GaP substrate)[7]. However, the electron wave function at the conduction band minimum has a special character[8]. It is expected to limit the performances of laser devices yielding high threshold current densities. Indeed, the conduction band of the GaAsP host material has a minimum at the X_{XY} point on the edge of the Brillouin zone, and partially localized electronic levels related to nitrogen incorporation lie at energies below this minimum. The conduction band minimum of GaNAsP/GaP QWs evidences a predominant localized N character[8]. Moreover, the maximum of the emission wavelength reported for such structures with reasonable N content is equal to 980 nm[7], which is not yet in the transparency window of Si.

Quantum dot (QD) lasers grown on GaAs or InP substrate display lower threshold currents due to the 0D density of states when compared with QW lasers on the same substrates[9]. (In,Ga)P QDs grown on GaP substrate have already been studied, and room temperature electroluminescence has been obtained[10]. However, theoretical studies have shown that the electronic band lineups correspond to a borderline case between type I and type II[11]. The (In,Ga)As(N)/GaP QDs system has recently attracted much attention. Fukami et al.[12] have claimed that the transparency window of silicon may be reached with InGaAsN/GaP QDs when In composition is 50% to approximately 60% and N composition is 1% to approximately 2%. In the following, InGaAs/GaP QDs are studied as a step toward InGaAsN/GaP QDs system. Both room-temperature photoluminescence (PL)[13] and electroluminescence[14] of InGaAs/GaP QDs have been recently reported. However, the description of the electronic band structure of this QD system is still lacking.

In this paper, we investigate (In,Ga)As/GaP QDs in a low-indium-content range both from the theoretical and experimental points of view. The effects of both indium composition and QD geometry is analyzed through a combination of **K·P** and tight-binding (TB) simulations. Optical properties are then studied by temperature-dependent photoluminescence and time-resolved photoluminescence (TRPL).

METHODS

(In,Ga)As QDs are grown on n-doped GaP(001) substrate using gas-source molecular beam epitaxy. After the growth of a 450-nm buffer layer and a 4-monolayer (ML) (In,Ga)As deposition with 30 s of annealing under As (see the work of Nguyen Thanh et al.[13] for more information), a 30-nm GaP capping layer is finally deposited to prevent surface non-radiative recombinations. The growth temperature is set at 580°C. The nominal composition of indium is set at 30%, but because of high growth temperature, indium effective composition is assumed to be below or equal to 15%.

Temperature-dependent PL experiments are carried out by exciting samples with a 405-nm continuous-wave laser diode. Power density is roughly estimated at 80 W·cm^{-2}. Samples are set in a helium bath closed-cycle cryostat to study PL from 10 K to room temperature. Measurements are also performed above room temperature using a hot plate. Attention is given to avoid the red luminescence of the deep centers in n-doped GaP substrates. Actually, the penetration length of the 405-nm beam is lower than the thickness of the GaP buffer layer, avoiding the substrate to be excited. Secondly, similar PL spectra are obtained on the same structures on non-doped GaP substrate, thus excluding any significant contribution from the GaP deep center luminescence.

For TRPL measurements, the sample is excited by a frequency-doubled Ti/sapphire laser at the wavelength of 405 nm. The repetition rate is 80 MHz. The PL signal is analyzed by an S20 streak camera, and measurements are performed at 10 K to overcome non-radiative recombinations channels.

RESULTS AND DISCUSSION

Band structure calculations

The eight-band **K·P** method has been extensively used to accurately simulate the electronic band structure of QDs with type I band alignment and direct optical transition (InAs/GaAs, InAs/InP...)[15, 16]. The case of InGaAs/GaP QDs in the low-In-content range is expected to be trickier because of the coupling of zone center conduction band states with conduction band states located on the edge of the Brillouin zone[13]. To deal with this issue, we simulate the direct optical transition with the eight-band **K·P** method. To get an estimation of X-like and L-like state energies in the dot, we consider the TB $sp^3d^5s^*$ model[17] for a QW with a thickness equal to the height of the dot. Thus, the lateral quantum confinement effect is disregarded but is assumed to have negligible effects on lateral valleys with large effective masses.

To consider realistic QD geometries for the simulation, the morphology of InGaAs/GaP QDs are imaged by plane-view scanning tunneling microscopy (STM). The 75 × 75-nm^2 STM image shown on Figure1a exhibits InGaAs/GaP QDs with approximately a cone shape. The in-plane anisotropic ratio (between length and width) is indeed measured in the range of 1 to 1.5. The statistical analysis of diameter and height distributions is presented in Figure1b. The **K·P** simulation is performed using the geometry defined on Figure2. A $C_{\infty v}$ symmetry is considered for QD geometries, and strain calculations are performed using elasticity and parameters of Vurgaftman et al.[18] and the finite element method for numerical computation. Three typical dimension sets representative of the inhomogeneous size distributions are summarized in the table of Figure2. The A, B, and C geometries correspond to real QDs typically found in the sample (see Figure1b). The D geometry is chosen to study theoretically larger QDs in order to address the problem of lowering the emission energy. A typical wetting layer of 1-ML thick is added in the model to account for the Stranski-Krastanov growth mode. Deformation potentials and Luttinger parameters used in the **K·P** model are those extracted from the TB calculation for bulk InGaAs and GaP[17]. The valence band offsets are taken from recent *ab initio* calculations[19].

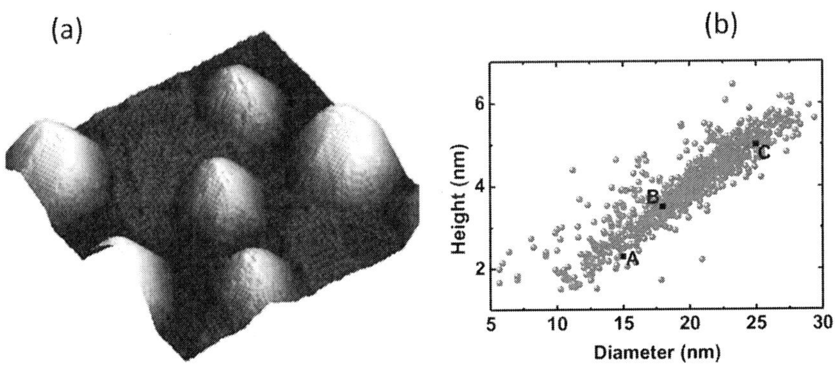

Figure 1. (IN,GA)AS QD IMAGE AND STATISTICAL CORRELATION. (**A**) A 75 × 75-nm² STM 3D plane view of (In,Ga)As QDs. (**B**) Statistical correlation between diameter and height on a 800 × 800-nm² image.

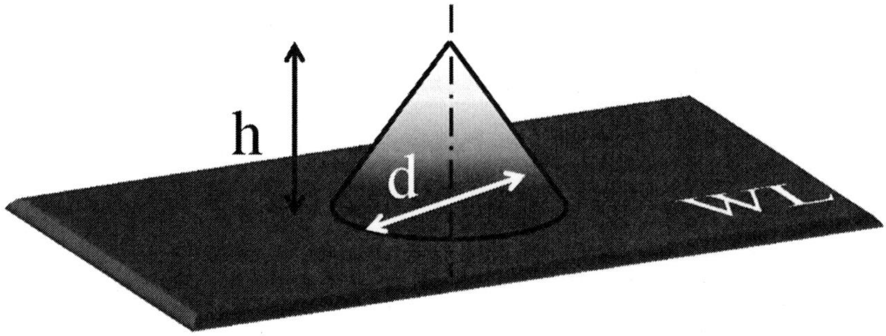

QD geometry	h (nm)	d (nm)
A	2.3	15
B	3.5	18
C	5	25
D	6.5	36

Figure 2. QD MORPHOLOGIES USED FOR THE EIGHT-BAND K·P CALCULATIONS.

The influence of In content is presented on Figure3a for a QD with the C geometry. The first electronic levels in the Γ, X, and L valleys and the first heavy-hole level of the QD are represented as a function of the In content. The electronic Γ and the heavy-hole levels are calculated with the **K·P** method, whereas the X and L electronic levels are calculated with the TB model. For low In content (below 30%), the ground optical transition is type I but is indirect with the first electron level of the X type. For very low In content (below 15%), the Γ-type conduction band level in the QD is even located at an energy above the one of the X-type conduction band of the GaP barrier. Nevertheless, a direct and type I ground state transition is predicted for In composition above 38%. This is a necessary condition in order to obtain a very efficient optical transition for such a QD geometry.

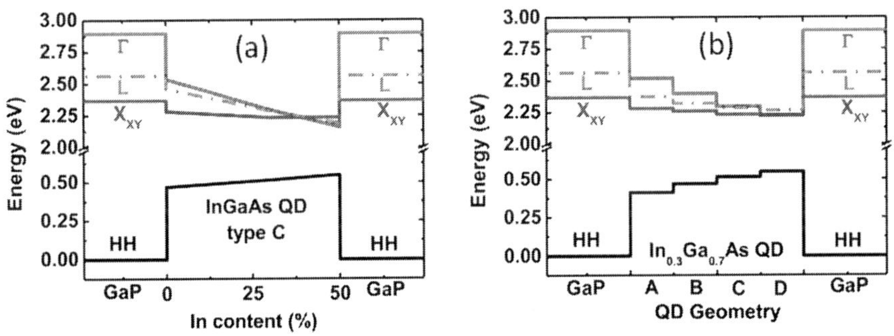

Figure 3. INFLUENCE OF IN CONTENT AND QD GEOMETRY ON THE ELECTRONIC LEVELS. (**A**) Electronic levels of $In_x Ga_{1-x}As$ QD with geometry C. (**B**) Electronic levels of $In_{0.3}Ga_{0.7}As$ QDs for the four geometries defined in Figure2. The Γ electronic level and the heavy-hole level in the QD are calculated with the k·p method. The X and L electronic levels are calculated with the TB model.

The influence of QD geometry is shown on Figure3b for a medium In content of 30%. For small QDs, the first Γ electronic level undergoes an important quantum confinement effect which lifts up this level above both X and L levels, which are less affected by confinement effects. An indirect to direct type crossover is predicted for large QDs.

In conclusion, an increase of In content and an enlargement of QD size are expected to lower the first Γ-type conduction band level and thus yield an efficient optical transition. Moreover, the indirect to direct type crossover should be induced by strain relaxation associated to In content increase and QD enlargement.

Optical properties

Continuous-wave PL spectra are presented in Figure4 for various temperatures. At 12 K, the PL spectrum exhibits a single peak centered at 1.78 eV. The peak shape undergoes a strong evolution from low to high temperatures. At 260 K, a shoulder appears on the high-energy side of the spectrum. At 300 K, a second optical transition clearly appears. When increasing the temperature above 300 K, the maximum of PL intensity switches from the low-energy (LE) transition to the higher-energy (HE) transition. This behavior may indicate that the HE optical transition is more efficient than the LE one. At 300 K, the LE transition is reported at 1.74 eV and the HE transition at 1.84 eV.

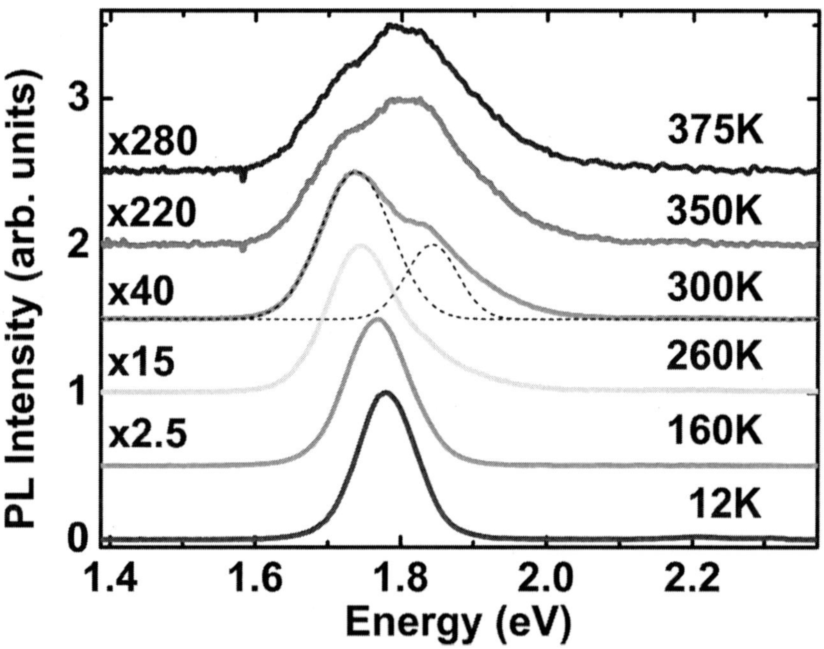

Figure 4. TEMPERATURE-DEPENDENT PL SPECTRA OF (IN,GA)AS/GAP QDS. The black thin dashed lines show the fit of the two transitions by two Gaussian peaks.

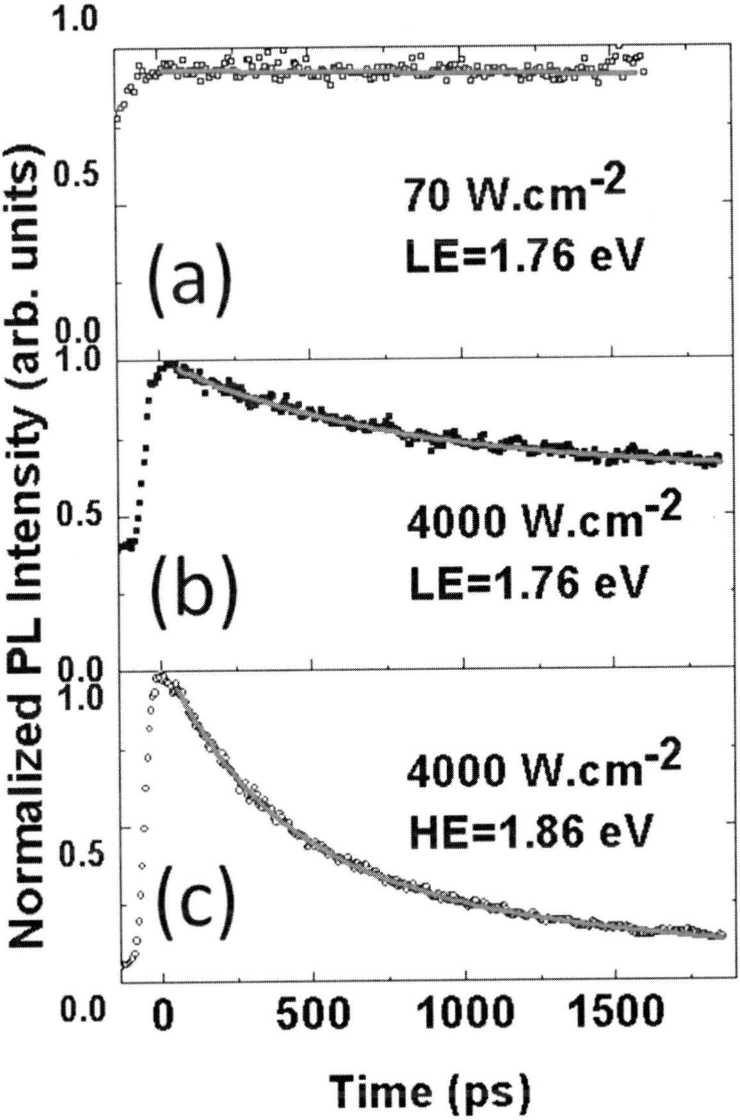

Figure 5. PL DYNAMICS AT 10 K AT SELECTED ENERGIES, LE AND HE. For power densities of (**A**) 70 and (**B, C**) 4,000 W·cm^{-2}. Red lines show biexponential fits.

To understand the nature of these two optical transitions, the dynamics of the recombination of carriers are investigated through TRPL spectroscopy. Experiments are performed at 10 K to overcome non-radiative recombination channels. The radiative lifetimes are deduced from the measured PL decay times. The sample is first excited with a low-incident power density equal to 70 W·cm^{-2}. The LE optical transition is only detected in accordance with the spectrum shown on Figure4 at low temperature. The evolution of the emission as a function of time is shown on Figure5a. The LE optical transition exhibits a very long decay time which is greater than the repetition period of the laser (12 ns) and is not easily measurable with this experimental setup. Such a long lifetime can be interpreted on the basis of the theoretical results of the previous section. The energy position of the LE PL peak at low temperature (E_{LE} = 1.78 eV) is consistent with the calculated indirect transition (between 1.74 and 1.79 eV) for medium-sized dots in the 0% to 15% In content range.

The sample is then excited with a large power density equal to 4,000 W·cm^{-2} in order to fill the low-energy electronic levels and allow the HE optical transition to occur. The PL dynamics at selected energies, E_{LE} and E_{HE}, are respectively shown on Figure5b,c. The time-resolved emission related to the LE transition can be fitted by the sum of a shorter exponential decay with a lifetime of 770 ps and a constant associated with the very long lifetime of the indirect transition. Many-body Auger effects leading to an enhancement of intradot carrier relaxation may lower the optical transition lifetime. The density of electron-hole pairs is indeed estimated to be high (above 10 per QD). Such effects have been observed in InAs/InP QDs[20]. For the HE transition, the emission shows a biexponential decay with short lifetimes of 340 and 1,700 ps, respectively. Both times are consistent with a direct type-I electronic transition in QDs and a better overlap of electron and hole wave functions. The $E_{HE}E_{LE}$ difference is also in reasonable agreement with that of theoretical calculations. For large-sized dots and In content of 15%, an energy difference of 100 meV is indeed calculated between both direct and indirect optical transitions.

CONCLUSIONS

The (In,Ga)As/GaP QD system is studied both theoretically and experimentally. The simulation results of **K·P** and TB methods are coupled and predict an indirect to direct crossover with the increase of In content and the ripening of QDs. Optical properties are then studied in the low-In-content range. In agreement with theoretical results, TRPL measurements are consistent with a ground optical transition of indirect type. A direct optical transition can be observed for high-power density or at room temperature where electrons get enough thermal energy to partially fill the Γ-type conduction band state.

ACKNOWLEDGMENTS

This research is supported by 'Région Bretagne' through the PONANT project including FEDER funds. This work was performed using HPC resources from GENCI CINES and IDRIS 2012-c2012096724. The work is also supported through the participation of the SINPHONIC JC JC ANR project N° 2011 JS03 006 and NANOTRANS C'NANO research program.

AUTHORS' CONTRIBUTIONS

CR performed the optical property measurements and theoretical calculations. TNT, CC, and NB performed the MBE growth and the analyses of quantum dot geometry. PT and ST performed the STM measurements. MP, AB, HF, and XM supervised the measurements of optical properties. LP, MON, JE, and JMJ developed the simulation of electronic properties. OD and ALC managed the team. All authors read and approved the final manuscript.

REFERENCES

1. Liang D, Bowers JE: Recent progress in lasers on silicon. Nat Photon 2010, 4: 511–517. 10.1038/nphoton.2010.167
2. Egawa T, Murata Y, Jimbo T, Umeno M: Low-temperature continuous-wave operation of AlGaAs-GaAs vertical-cavity surface-emitting lasers on Si substrates. Photonics Technology Letters, IEEE 1997, 9: 872–874.
3. Momose K, Yonezu H, Fujimoto Y, Furukawa Y, Motomura Y, Aiki K: Dislocation-free and lattice-matched Si/$GaP_{1-x}N_x$/Si structure for photo-electronic integrated systems. Appl Phys Lett 2001, 79: 4151. 10.1063/1.1425451
4. Volz K, Beyer A, Witte W, Ohlmann J, Németh I, Kunert B, Stolz W: GaP-nucleation on exact Si (0 0 1) substrates for III/V device integration. J Cryst Growth 2011, 315: 37–47. 10.1016/j.jcrysgro.2010.10.036
5. Létoublon A, Guo W, Cornet C, Boulle A, Véron M, Bondi A, Durand O, Rohel T, Dehaese O, Chevalier N, Bertru N, Le Corre A: X-ray study of antiphase domains and their stability in MBE grown GaP on Si. J Cryst Growth 2011, 323: 409–412. 10.1016/j.jcrysgro.2010.10.137
6. Liebich S, Zimprich M, Beyer A, Lange C, Franzbach DJ, Chatterjee S, Hossain N, Sweeney SJ, Volz K, Kunert B, Stolz W: Laser operation of Ga(NAsP) lattice-matched to (001) silicon substrate. Appl Phys Lett 2011, 99: 071109. 10.1063/1.3624927
7. Hossain N, Sweeney SJ, Rogowsky S, Ostendorf R, Wagner J, Liebich S, Zimprich M, Volz K, Kunert B, Stolz W: Reduced threshold current dilute nitride Ga(NAsP)/GaP quantum well lasers grown by MOVPE. Electron Lett 2011, 47: 931–933. 10.1049/el.2011.1927
8. Robert C, Perrin M, Cornet C, Even J, Jancu JM: Atomistic calculations of Ga(NAsP)/GaP(N) quantum wells on silicon substrate: band structure and optical gain. Appl Phys Lett 2012, 100: 111901. 10.1063/1.3694028

9. Ledentsov NN, Grundmann M, Heinrichsdorff F, Bimberg D, Ustinov VM, Zhukov AE, Maximov MV, Alferov ZI, Lott JA: Quantum-dot heterostructure lasers. IEEE Journal of Selected Topics in Quantum Electronics 2000, 6: 439–451.

10. Hatami F, Lordi V, Harris JS, Kostial H, Masselink WT: Red light-emitting diodes based on InP/GaP quantum dots. J Appl Phys 2005, 97: 096106. 10.1063/1.1884752

11. Hatami F, Masselink WT, Schrottke L, Tomm JW, Talalaev V, Kristukat C, Goñi AR: InP quantum dots embedded in GaP: optical properties and carrier dynamics. Phys Rev B 2003, 67: 085306.

12. Fukami F, Umeno K, Furukawa Y, Urakami N, Mitsuyoshi S, Okada H, Yonezu H, Wakahara A: Analysis of quantum levels for self-assembled InGaAsN/GaP quantum dots. Physica Status Solidi (c) 2011, 8: 322–324. 10.1002/pssc.201000500

13. Nguyen Thanh T, Robert C, Cornet C, Perrin M, Jancu JM, Bertru N, Even J, Chevalier N, Folliot H, Durand O, Le Corre A: Room temperature photoluminescence of high density (In,Ga)As/GaP quantum dots. Appl Phys Lett 2011, 99: 143123. 10.1063/1.3646911

14. Song Y, Larry Lee M: Room temperature electroluminescence from light-emitting diodes based on $In_{0.5}Ga_{0.5}As$/GaP self-assembled quantum dots. Appl Phys Lett 2012, 100: 251904. 10.1063/1.4729591

15. Stier O, Grundmann M, Bimberg D: Electronic and optical properties of strained quantum dots modeled by 8-band k·p theory. Phys Rev B 1999, 59: 5688–5701. 10.1103/PhysRevB.59.5688

16. Cornet C, Schliwa A, Even J, Doré F, Celebi C, Létoublon A, Macé E, Paranthoën C, Simon A, Koenraad PM, Bertru N, Bimberg D, Loualiche S: Electronic and optical properties of InAs/InP quantum dots on InP(100) and InP(311)B substrates: theory and experiment. Phys Rev B 2006, 74: 035312.

17. Jancu J-M, Scholz R, Beltram F, Bassani F: Empirical spds* tight-binding calculation for cubic semiconductors: general method and material parameters. Phys Rev B 1998, 57: 6493.

18. Vurgaftman I, Meyer JR, Ram-Mohan LR: Band parameters for III-V compound semiconductors and their alloys. J Appl Phys 2001, 89: 5815. 10.1063/1.1368156

19. Li Y-H, Walsh A, Chen S, Yin W-J, Yang J-H, Li J, Da Silva JLF, Gong XG, Wei S-H: Revised ab initio natural band offsets of all group IV, II-VI, and III-V semiconductors. Appl Phys Lett 2009, 94: 212109. 10.1063/1.3143626

20. Cornet C, Labbé C, Folliot H, Caroff P, Levallois C, Dehaese O, Even J, Le Corre A, Loualiche S: Time-resolved pump probe of 1.55 μm InAs/InP quantum dots under high resonant excitation. Appl Phys Lett 2006, 88: 171502. 10.1063/1.2199454

5 CHAPTER

QUANTUM DOTS PREPARED BY DROPLET EPITAXIAL METHOD

Ákos Nemcsics[1, 2]

[1] *Institute for Microelectronics and Technology, Óbuda University Budapest, Hungary*

[2] *CER, Institute for Technical Physics and Materials Science, Hungarian Academy of Sciences, Budapest, Hungary*

1. INTRODUCTION

The electrons "boxed up" or confined in quantum confinement and the calculation of their energy levels, which are well known from textbooks, were regarded for decades as idealistic concepts far from practical reality. The molecular beam epitaxial (MBE) technology, developed for compound semiconductors, made the introduction of the low-dimension structures into the everyday scientific research possible. The structures confined in all three dimensions are called quantum dots (QDs).

For the fabrication of QDs or any zero-dimensional structures, various methods were developed. For a long time, the only known method for the production of epitaxially grown zero-dimensional system was the strain-induced method, based on lattice mismatch. Presently, the most used technique is the lattice-mismatched technique by the Stranski–Krastanov mode [1–5]. QDs, such as InAs nanocluster prepared on GaAs surface, are the oldest known system. The droplet epitaxy (DE), however, is one of the latest methods for QD preparation. This method of preparation is not only an alternative way to the conventional method but also a production method for a number of zero-dimensional quantum structures such as ringlike QDs, double ringlike QDs, inverse QDs (filled nanoholes), or QD molecules. This technique is also fully compatible with MBE technology.

The subjects of this paper are the technologies used, growth kinetics, and some properties of the zero-dimensional nanostructures of different shapes, grown by DE. The fundamentals of DE were first demonstrated by Koguchi and his coworkers in the early 1990s [6–9]. One advantage is that it is not restricted by mismatch conditions; in addition to that is the possible fabrication of strain-free QDs and similar nanostructures by DE, against the Stranski–Krastanov-based growth method. The shape diversity makes it a preferable process for that application. The size, shape, and elementary distribution of the developed structures are dependent on the developing technology used. It is obvious that the physical parameters of the structures are important from both the theoretical and practical point of view.

In this paper, QDs grown by DE will be described, selected from III and V class materials. Here, the QDs are mostly GaAs with the use of AlGaAs as barrier material. This process consists of the following two basic steps: first, the metallic (e.g., Ga) nanosized droplets are generated on the surface by the Volmer–Weber growth mode, followed by the second step of droplet crystallization that is the transformation into GaAs QDs in arsenic atmosphere.

For the control of the process, the kinetics of the growth process knowledge is necessary, which is so far lacking the full theoretical description.

The DE technology for QD preparation has superiority over the nowadays widely used strain-induced method. The greatest advantage is that both lattice matched and mismatched can be used for QD fabrication. At the same time, the size and density of the QD's can be controlled independently, and the distribution is more uniform. However, the shapes of the QDs show a wide spectrum. This spectrum includes the conventionally shaped QDs, ringlike QDs, and complex multicomponent QD structures [9–13]. To avoid intermixing, low temperature can be applied at preparation. A wide variety substrate orientation has beneficial effect on the application.

Although the technique is quite recent, it is already successfully applied in a number of cases. QDs had significantly improved the general performance of the optoelectronic devices, like semiconductor lasers and optical amplifiers. DE fabricated QDs introduced further improvements for the beneficial reasons described previously [10].

Future quantum devices, based on quantum mechanical and electromagnetic interactions, require lateral QD configurations. With the modified DE method, aligned QD pairs and QD molecules can be produced without involving lithographic technology [14, 15].

2. THE DROPLET EPITAXIAL TECHNIQUE

2.1. Technological background and preliminaries of the droplet epitaxial technique

It is well known in our profession that the grown structure can be classified in three growth modes depending on the interaction among the constituents. These modes are the layer-by-layer growth or Frank–van der Merve mode, the island growth or Volmer–Weber mode, and the intermediate mode (layer-plus-island growth) or Stranski–Krastanov mode (Figure 1) [16]. The morphology of the interface is determined by the interplay between deposition, desorption, and surface diffusion. If an atom or molecule from the environment arrives at a random position on the surface, the deposition process bonds with the surface atoms and sticks.

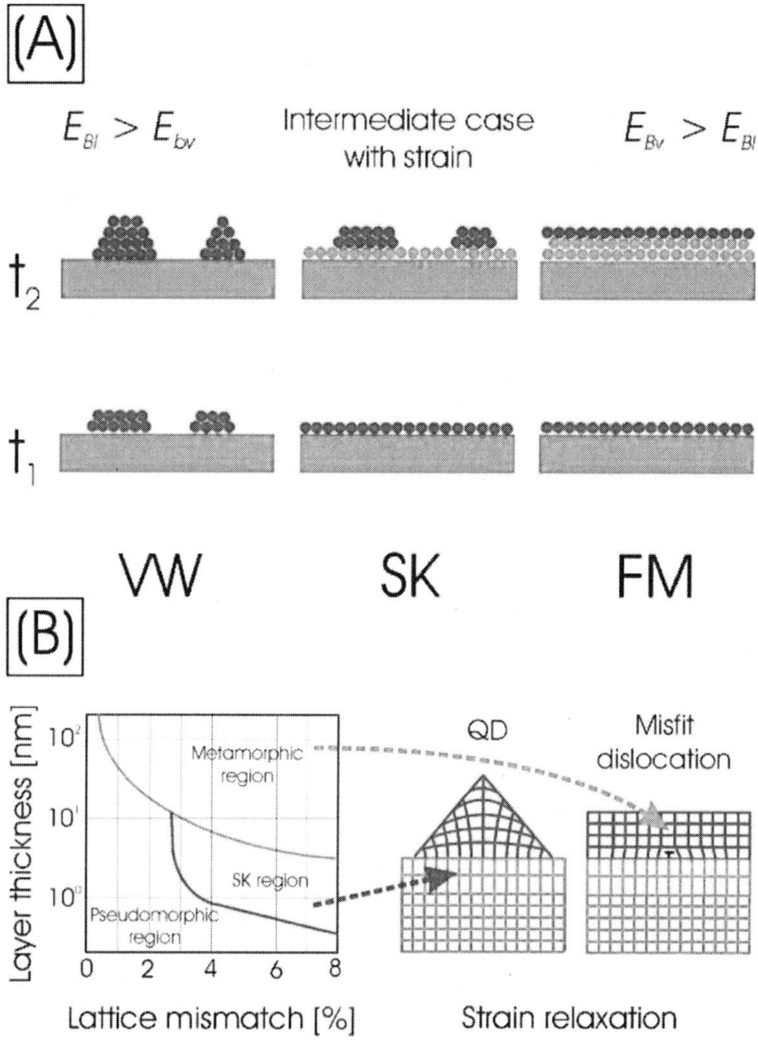

Figure 1. (A) Schematic illustration of the three growth modes. (B) Strain relaxation via QD formation and via generation of misfit dislocation.

The process competing with deposition is desorption. During the desorption process, some atoms deposited on the surface leave the interface. The probability of the desorption depends on how strongly the atom is bonded to the surface. The strength of the bonds depends either on the type of atom or on the local atom arrangement of the surface where the atom sticks. The desorption can be negligible for many materials under typical MBE conditions. The deposited atoms diffuse on the crystal surface, searching for the energetically most favorable position. The diffusion length can be quite large and depends on the temperature and the binding energies to the substrate (E_{Bv}, binding in vertical direction).

In the Frank–van der Merve mode, the interaction between the neighboring atoms (E_{Bl}, lateral directions) in the layer is weaker than that with the substrate atoms ($E_{Bv} > E_{Bl}$) [17]. In this growth mode, islands of monolayer (ML) height coalesce before a new layer can nucleate on top of them. In the Volmer–Weber mode, the situation is just the opposite, that is, the binding energy among the deposited atoms is stronger than to the substrate surface ($E_{Bl} > E_{Bv}$). In this case, the growth proceeds to many atomic layers at discrete islands before these islands merge.

In the Stranski–Krastanov mode, the process is initiated in a layer-by-layer fashion, but islanding commences after the growth of a certain layer thickness. In certain situations, layer-by-layer growth is desirable because of the need for multilayered structures with exact layer thickness and flat interfaces. This requires that the nucleation takes place as a single event on the substrate. In this growth mode, the lattice mismatch between the grown layer and the substrate material is the most common case.

The deformation takes place by the even layer to match the lattice of the substrate. In this case, elastic energy is generated in the structure, which deforms mainly the lattice of the grown layer. When this deformation energy overcomes a critical value, it can be relaxed in two ways. One possible way is to nucleate nanosized islands on the surface of the even layer, which is called the wetting layer. The absence of the lateral layer allows the atomic planes to laterally relax, reducing the elastic energy in the structure. According the first mentioned way, the so-called strain-induced QD can be grown. The other way is to generate misfit dislocations at the interface [18–21]. The dislocations in the substrate and at the interface can overgrow into the grown layer [23–25]. These defects are called threading dislocations.

Figure 1 shows the following two out of three growth modes, Volmer–Weber and Stranski–Krastanov for nanostructure production. One may ask why only the Stranski–Krastanov growth mode is widely used for QD production. The answer lies in the development of the MBE technology. The application of III–V materials opened the way for the production of heteroepitaxial structures, specifically the GaAs–AlGaAs–AlAs system, for its identical lattice parameter. The growth of InAs and InGaAs created particular technological difficulties due to the lattice mismatch. In the early 1990s, the success of InGaAs growth on GaAs, with differing indium content [26–27], brought about a big advance in this field. Initially, the thickness of the layers stayed under a critical value, but this limitation was successfully overcome. The strain-induced QD production has superseded the DE, dominating the field earlier.

The archetypal system of the clustered nanostructures is the lattice-mismatched system such as InAs on GaAs, where the strain-induced process leads to the formation of QD. The lattice-mismatched structure can relax either by the generation of misfit dislocation or by the nucleation of nanosized islands on the even layer (Figure 1). The essential driving force for coherent QD formation, after an under layer formed, is the strain relaxation, whereby the energy gained from the increase in surface area via QD formation more than compensates the increase in interfacial free energy. It is generally accepted that one of the essential driving forces, for coherent lattice-mismatched QD formation, is strain relaxation.

The field of self-organized strain-induced QDs is dominated by two material system, InAs/GaAs (001) and Ge/Si (001). For instance, the MBE growth of about 1.5 MLs of InAs on GaAs results in the spontaneous formation of InAs QDs driven by the strain between the deposited InAs and the GaAs substrate. The driving force for the self-organized QD formation is the strain energy induced by the lattice mismatch of about 7%, in which the condition restricts the material choice. Two families of the shape such as pyramids and domes can be created during the defect-free QD transformations [28].

2.2. Principles of DE

The QD preparation in a self-assembled manner by using the DE technique is an advantageous alternative and extension to the strain-induced QD creation. The basic idea of the DE originated from Koguchi and his coworkers [29, 30]. In comparison with the technology of the strain-induced Stranski–Krastanov QD production, the DE is more flexible regarding the choice of the QD material and also regarding the shape and distribution of the resulted QDs and other unconventionally-shaped QD structures. For example, it is possible to create not only InAs QD on GaAs but also strain-free GaAs QD on AlGaAs surface. Furthermore, this technology of DE allows us to produce ringlike QD and further special-shaped nanosized structures and structure complexes [31–35].

In the case of the DE, the clustering on the surface is carried out with the help of Volmer–Weber like growth mode. The basic idea based on the separation of the III-column and V-column materials supply during the MBE growth (Figure 2). Here, the QD preparation consists of two main parts such as the formation of metallic nanosized droplet on the surface and its crystallization with the help of the non-metallic component of the compound semiconductor. It is very important to mention that this DE technique is entirely compatible with the MBE technology. This circumstance allows us to combine the DE method with the other conventional MBE processes.

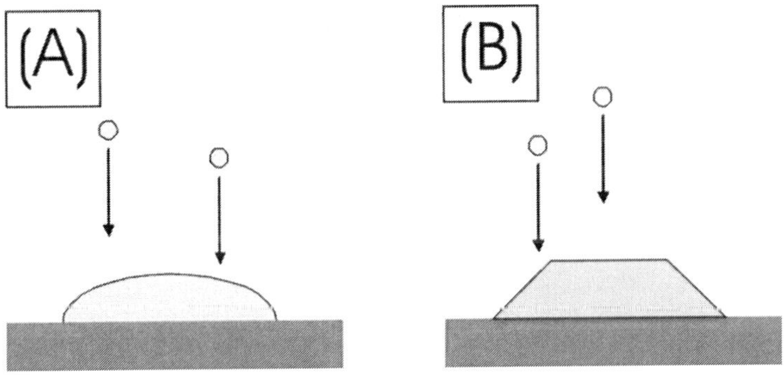

Figure 2. (A) First step of DE: generation of metallic droplet on the surface. (B) Second step: crystallization to create QD.

First, the III-column elements such as Ga, In, Al, etc., are deposited. These metallic components form clusters that are nanosized droplets on the substrate surface. At the droplet formation, the main driving forces are the diffusion, the minimalization of the surface tension, and the Ostwald ripening [36–42]. The dominance of these driving forces depends on the technological circumstances. The second step of the DE technique is the crystallization of the metallic cluster by the molecular beam irradiation of the nonmetallic V-column element. It is important to suppress or to control the lateral diffusion of the group III elements from the cluster.

2.3. QD engineering

As known, the electronic and optoelectronic properties of the nanostructure depend strongly on the shape of the nanostructure. The technology used for growth can be controlled by the size, shape, and elementary distribution of the developing structure. These technological and physical parameters are very important from both theoretical and practical points of view. In the DE technique, because of the shape and distribution diversity, the key point is to discover growth mechanism and its relation with the parameters of the developed structures. Since the discovery of the DE, several scientific knowledge are gathered. However, until now, no full theoretical description of the underlying growth kinetic is available.

An interesting correlation between the GaAs QD shapes and their volume was observed by Heyn and coworkers [43, 44]. As a result of this discovery, they developed the first growth model for the DE of GaAs QDs [43, 44]. The experiment was carried out with the cystallization of Ga droplets on AlGaAs surface. Here, growth temperature ranged between 140°C and 300°C. On the arsenic terminated surface, the deposition of Ga with flux 0.025–0.79 ML/s was executed. It was clearly visible that growth temperature strongly influenced QD density.

The QD density dependence on temperature shows the scaling law [44]. Under 200°C, the slope of the temperature shows E_a = 0.235 eV.

Over 200°C, data did not follow the scaling law, but QD density decreased. At this temperature, coarsening of the cluster begins. This effect goes back to the Ostwald ripening, which means the growth of large clusters at the expense of smaller ones, hence causing a decrease in the total cluster density.

Higher temperature combined with lover arsenic flux will result in a ringlike nanostructure. At longer waiting time and lower arsenic flux, these rings fall below the original substrate surface and the encircling surround. After the heat treatment of the sample with the clusters formed, the total lack of arsenic results in nanoholes. The creation of nanoholes can lead back to the thermal etching of the substrate by the liquid metallic component. At a typical process temperature of 570°C, etching starts approximately 20 s later than the beginning of the annealing step, followed by desorption approximately 100 s after. The depth of the nanoholes can be found by multiplying the etching time by the etching rate.

These nanoholes can be filled with low band-gap material to create the so-called inverted QD. Furthermore, the nanoholes can serve as templates for QD preparation. In the utilization of surface anisotropy, QD pairs can be grown. The nanohole and the nanomound can initialize QD molecules, where the QD per molecules ranged from two to six. In the case of InAs QDs on GaAs (001) surface, the number of QDs per GaAs mound can be effectively controlled by varying InAs ML coverage [45].

3. DROPLET FORMATION

3.1. The initial surface status

During the QD formation, the status of the substrate surface is very important. It is well known that surface reconstruction influences strongly the growing layer and the forming structure [16]. Surface reconstruction depends on the temperature and the ambient environment environment, too. As mentioned before, the DE technique is compatible with the MBE technology. It means that the growing surface can be continuously monitored with reflection high-energy electron diffraction (RHEED) to determine the status of the surface, e.g., the surface reconstruction. During the growth, the RHEED pattern and its specular spot intensity allows a real-time tracking of the surface status changes. Here, we are using the technological most important (001) surface orientation. The bipolar (001) surface of GaAs can be terminated by either Ga or As atoms with rich variety of reconstructions are formed depending on surface stoichiometries [16].

Before the droplet creation, we have to grow a layer from a wide band-gap material for barrier. This layer growth happened with a conventional MBE technology, which is carried out in arsenic ambient pressure. Hence, the initial status of the surface is arsenic terminated. After the layer growth, we have to change the temperature depending on the planned nanostructure shape. The deposited MLs are not equal with the droplet volume. Assume that Ga is deposited on AlGaAs surface at 200°C. Due to the strong binding energy of As to Ga and substrate, the first Ga ML is consumed for the formation of Ga terminated surface. This first ML volume does not contribute to the formation of the Ga droplets. It means the coverage of Ga located in the droplet is $Ft\text{-}A$. The total Ga surface coverage is $\Theta = Ft$, where it is resulted with flux F for a time t, and A is the incorporated value of ML.

Although no complete phase diagram of GaAs (001) has been mapped, certain surface phases are generally observed during growth or after growth and annealing. For example, the less As-rich surface exhibits a (2×4) / $c(2 \times 8)$ pattern and arises from 0.75 ML of As. This reconstruction is normally present during the MBE of GaAs. Areas with (2×4) and $c(2 \times 8)$ symmetry can coexist. With larger coverage of As, a $c(4 \times 4)$ reconstruction with As coverage of 1.75 ML is observed [46, 47]. Depending on the substrate temperature and the Ga-to-As ratio during MBE, the GaAs (001) surface is known to display various surface reconstructions ranging from the As-rich $c(4 \times 4)\beta$, $c(4 \times 4)\alpha$, $c(2 \times 8)$, and (2×4) to the Ga-rich (6×6), $c(8 \times 2)$, (4×6), and (4×2) surface. The surface symmetry critically depends on the preparation conditions [46, 47].

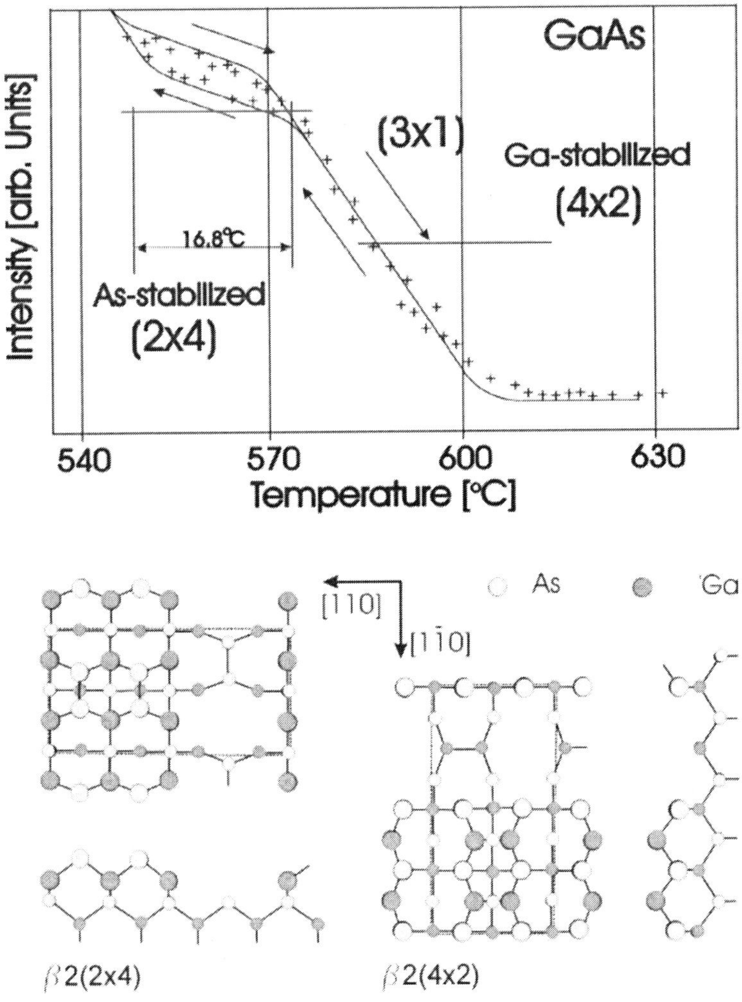

Figure 3. Ga and As stabilized GaAs (001) surface depending on the substrate temperature.

In situ investigations of this large set of reconstructions with RHEED are well established. Using scanning tunneling microscopy (STM), most of these surface reconstructions could also be imaged in real space, and the structure and the structural models were refined, too. Most of the previous studies were based on the diffraction techniques and no direct observation for the (4 × 6) phase exists. The STM study reveals that (4 × 6) symmetry arises from the coexistence of (2 × 4) and (4 × 2) units [48]. We have seen before that the change in temperature and other parameters results in the continuous change in surface reconstruction. During the change of the temperature, not only the reconstruction but also the specular spot intensity changes. The results show that at lower temperature, the specular spot intensity is high, and at higher temperature, the intensity is low. Furthermore, in the up and down and back directions, the temperature versus intensity function shows hysteretic behavior (Figure 3). The explanation of this phenomenon can lead us to the understanding of the intricate surface structures [49, 50].

The initial step of the DE is the droplets creation on the surface. It is already well documented that the droplet nucleation requires Ga-terminated surface, which means that a part of Ga coverage incorporate into the surface. During the initial stage of Ga deposition, fundamental change happens in the surface reconstruction, which strongly influences the droplet formation. Before the Ga deposition, the surface reconstruction can be usually *c*(4 × 4), (2 × 4), or (4 × 6), where the values of As coverage are 1.75, 0.75, or 1.12 ML, respectively. The previous paragraphs show the complicity of the surface reconstruction and other difficult unsolved problems associated with the process. The part from the metallic surface deposition migrating into the droplet depends largely on the initial state of the surface.

3.2. Size and density of the droplets

The size and the density of the metallic droplets depend on the substrate temperature and on the deposited MLs. In the case of Ga droplets on AlGaAs surface under 200°C, the cluster density versus temperature follows scaling law. Over 200°C, the scaling law is broken. Under this temperature, the clusters begin to coalesce, which drastically reduce the cluster number. This decrease of cluster number is governed by the Ostwald ripening [36–42].

The reduction of the cluster number is affected by the diffusion length of the constituents. The condition required for the coalescence is the overlap of the diffusion length and the mean distance of the neighboring clusters. It characterized by a local interaction between neighboring clusters of slightly different size. The smaller cluster starts to decompose to maintain the gradient of the radius equilibrium concentration toward the larger cluster due to the Gibbs–Thomson effect [51]. The appearance of such processes is critically dependent on the initial conditions.

The Ostwald ripening describes the driving force, which determines the size distribution of cluster ensemble on the surface [38–40, 42, 52]. This description supposes a driving force that depends on the mean radius of the islands. It is supposed that atoms detach more readily from smaller islands and condense rather to larger ones. Therefore, for a given amount of material on the surface, larger islands grow at the expense of smaller ones. The temporal evolution of the island radius (r) can be described by a differential equation: $dr/dt = b(T) / r^2(1/r - 1/r_c)$, where r_c is the critical cluster radius and $b(T)$ is a temperature-dependent parameter.

3.3. Droplet–surface interaction

In this section, we will discuss the interaction between the droplet and the substrate surface. Here, the dependence of the droplet shape and the so-called thermal etching are discussed. It is known that the droplet shape is determined by the wettability of the substrate surface, and this property can be described by contact angle. There are three types of situations: no wetting, partial wetting, and complete wetting, where these situations correspond somewhat with the three growth modes. The contact angle depends on some circumstances, such as the status of the carrier surface, the droplet size, and the temperature. This is because the nuclei may wet the substrate, strongly changing their geometry. The deposited material will not wet the substrate because this would be accompanied by an overall increase in free energy of the system. The liquid deposit will spread out to maximize the area of the interface. For all the other situations, a partial wetting with a contact angle θ should be considered, that is, $\gamma_s = \gamma_i + \gamma_e \cos\theta$, where γ_e and γ_s represent the surface free energies of deposited cluster and substrate, respectively, and γ_i is the interfacial free energy (Figure 4) [53].

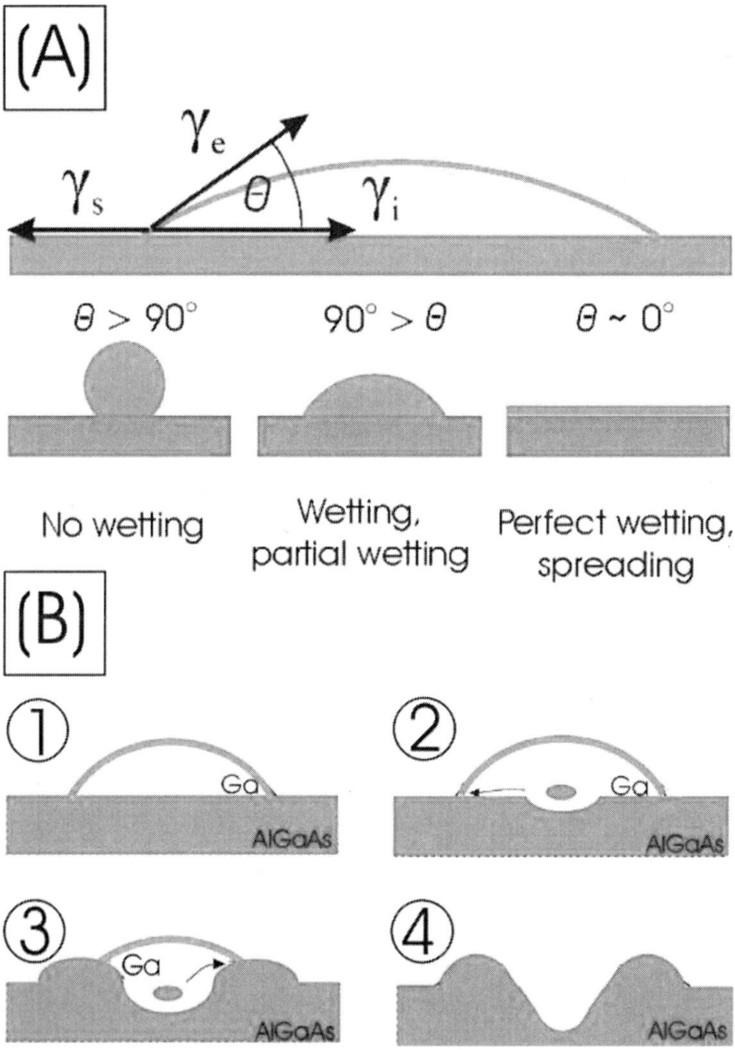

Figure 4. (A) The contact angle dependence on the surface tension and wettability. (B) Localized thermal etching of the compound semiconductor surface by a metallic droplet.

In our DE case, we have to discuss a further interaction between the droplet and the carrier surface. This interaction is called thermal etching, droplet etching, or local droplet etching. The explanation of the process will be discussed with the help of Ga/GaAs/AlGaAs system. Where Ga droplet forms on AlGaAs surface.

It is known from the liquid phase epitaxy that thermal etching takes place at the interface Ga melt and AlGaAs surface [54–57]. In the first step in DE technology, Ga droplets are generated on the surface. Due to thermal etching, this droplet-shaped Ga melt can solve the arsenide from the substrate material. At the interface, the metallic components of the substrate mix with the Ga atoms of the droplet. During the QD formation, this thermal etching takes place, too. In the initial stage of the QD formation, enough time is available for the mixing of the atoms at the interface. In the case of GaAs QD on AlGaAs substrate, Al content can be observed in the GaAs QD. This phenomenon was confirmed by the author and coworkers [58]. This thermal etching effect can be observed also at evolution of the ringlike QDs.

Our observations show that the middle part of ringlike QDs is below the surface of the substrate surrounding the QDs. This phenomenon is also the consequence of thermal etching. At slow crystallization, etching can act longer; therefore, it results in deeper holes in the center. At high temperature, quite deep nanohole can be achieved [85].

Thermal etching can be utilized also for nanostructure preparation. With the help of droplets created on the substrate surface, the surface can locally be etched. At higher temperature, this etching phenomenon is more significant. After the local thermal etching, the remained deep craters are the so-called nanoholes. During the DE QD fabrication, the Ga deposition temperature is ranged between 100°C and 350°C typically. These temperatures are very low compared to usual MBE growth conditions. In this case, the droplet thermal etching is less significant. At significant higher temperature, the situation is different.

If the temperature ranged between 450°C and 620°C, the local thermal etching under the droplet is already considerable. After an annealing time, the droplet is replaced in a nanohole surrounded with lobes. This local droplet etching shows a very promising technique for the structuring of the substrate surface without any lithographic processes. Furthermore, this patterning is also fully compatible with the MBE technology. In this way, nanohole was first presented by Wang and coworkers [59].

This finding opened a new way for the fabrication of another type of QD. This system was based on the filling of the nanohole, where the fill of the hole is carried out with lower band-gap material than the substrate. In this way created, nanohole filled inverted QDs are demonstrated by Heyn and coworkers [60–66]. In the process of nanohole filling, we utilize the Ga diffusion on the substrate surface. The temperature and orientation dependence of the Ga (and other metallic component) diffusion can be utilized at further nanostructure preparation.

4. FORMATION OF QD

4.1. Preparation technology

The initialization of the alternatively grown conventionally shaped QD preparation, the basic idea of DE, originated from Koguchi and his coworkers [6, 30]. In this way, it is possible to create strain-free GaAs QD on AlGaAs surface. Along this discovery, further unconventionally shaped QD types were discovered [9, 32–35]. In the following, we will detail the DE-grown single hump-shaped QD in the case of GaAs/AlGaAs system. As discussed above, this process consists roughly of two main steps. First, metal (in our case Ga) droplets are generated on the surface in a Volmer–Weber-like growth mode. In the second step, crystallization takes place. Under arsenic pressure, the droplets transform into crystalline nanostructures. The shape of the nanostructures and their distribution depend on the initial droplet size and distribution and on the further technological parameters such as substrate temperature, arsenic pressure, waiting time, etc.

Here, typical experimental parameters for conventional QD preparation are described [43]. The GaAs QDs are grown on AlGaAs (001) surface. The growth experiments are performed in a solid source MBE system equipped with effusion cells for Ga and Al evaporation and valved cracker cell for arsenic ambient pressure. The evolution of growth front is in situ monitored with RHEED. First, on GaAs (001) wafer, pure GaAs layer is grown, and it is followed by an AlGaAs layer with Al content of 0.3. After the AlGaAs layer preparation, the sample is cooled down to 200°C. The θ = 3.75 ML Ga is deposited with the flux of 0.75 ML/s without any As flux. After the Ga deposition, a 60-s waiting time is performed. The annealing is carried out at a temperature of 350°C and at an As pressure of 5×10^{-5} Torr. The growth process are tracked continuously in the direction of [11 ¯ 1¯ 0] with the help of RHEED [67]. In this direction, the RHEED pattern is more informative for the status of the QD evolution compared with the perpendicular [110] direction. After the growth process, the quantum objects can be ex situ investigated (AFM and TEM and PL methods), which can provide us further information to understand the growth kinetics.

During the QD preparation, the sequence of the RHEED pictures in the main growth stages is depicted in Figure 5. In the initial stage, the RHEED pattern showed sharp streaks (stage at t_0). After the Ga deposition, the pattern diffused on the RHEED screen (stage at t_1). Almost at the same time with the offering of arsenic pressure, the RHEED pattern changed suddenly from diffused to spotty (stage at t_2). During the annealing phase, the pattern changed slowly (some minutes) from spotty to spots with tails (stage at t_3). The density of QDs is 3.6×10^{10} cm^{-2}. It is shown that the characteristic RHEED pattern of QD is still recognizable even if the dot density is one order of magnitude less [43].

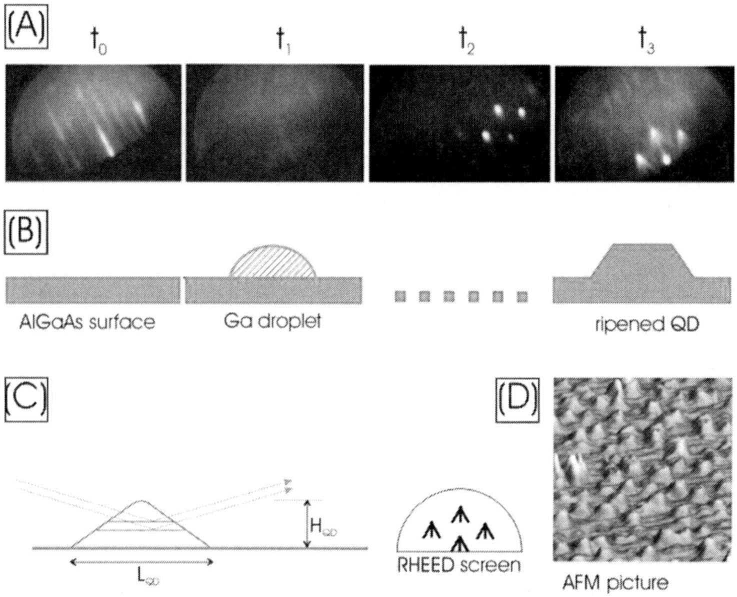

Figure 5. (A) During the QD evolution, the changing of the RHEED pattern and (B) its temporal evaluation. (C) Interaction of QD and electron beam. (D) AFM picture of DE-grown QDs.

4.2. Geometry and electronic structure of QD

As we demonstrated earlier, the shape of the QDs depends on the volume and also the density of QDs. According the shape, there are two types of QDs. The larger type of QD is a truncated pyramid-like shape with side facets of 55°. The smaller QD is a pyramid-like form with a side facet of 25° [43, 44]. In the following, the geometry of the QD will be discussed in the case of the smaller type of QD.

The atomic resolution structure of a typical DE-grown QD is shown in Figure 6. (The image was taken with the electron beam parallel to the [110] zone axis of the AlGaAs single crystal substrate [43, 67–68].) The typical dimensions of the QD shown in Figure 6 are 54 nm base width and 5.5 nm height. The steepness of the QD side wall measured by TEM as well as by AFM shows about 25°. Figure 6 also shows that all the lattice fringes of the AlGaAs substrate are continued in the GaAs QD without any distortion. It is well known that the lattice parameters of these two substances are practically the same. No crystal defects were observed within the QD or at the interface with the host crystal [68, 69].

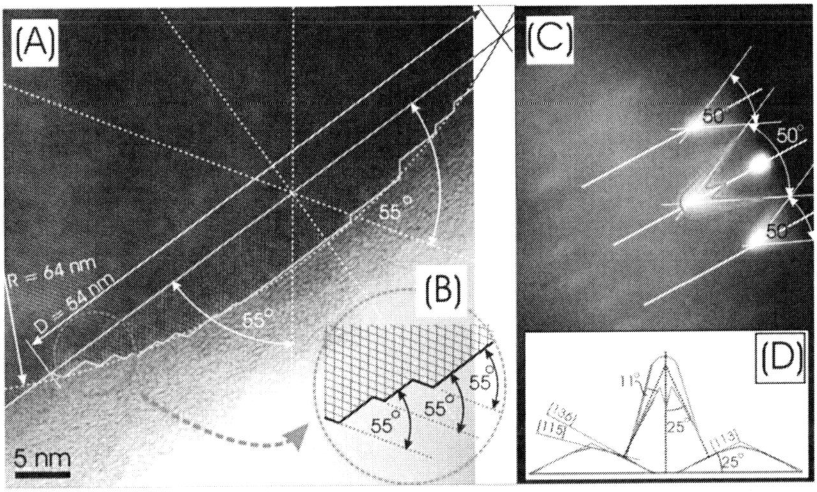

Figure 6. (A) Cross-sectional TEM image of a DE-grown QD, and (B) its stepped surface. (C) Broadened chevron tails and (D) their explanation.

The feature mentioned is shown in a darker contrast of the QD, also a few atomic surface layer of AlGaAs crystal between the QDs. The darker contrast on the AlGaAs surface can be interpreted as follows. The AlGaAs layer, at low temperature, shows As-rich $c(4\times4)$ surface. Nominally, θ = 3.75 ML Ga is deposited on the surface without arsenic flux. In the duration the Ga supply, a few ML of Ga is combined with excess arsenic surface atoms. Thus, a thin GaAs layer forms on the surface while the rest of the Ga forms nanodroplets. During the crystallization, the droplet and the surface layer go into similar composition. As a result, the QD and the surface layer shows similar darker contrast compared to the host material [69]. The existence of this GaAs surface layer was predicted earlier from the comparison of the PL measurement and the energy level calculation [32].

The chevron tails are connected with the faceting of QDs as it was verified in the case of diverse-shaped DE QDs [10]. In case of DE, it can be shown from the AFM measurement and from the tilted TEM picture that the shape and size of QDs are very uniform [69], and their side angle is about 25° [43, 68], which corresponds to the half opening angle of the chevron tails (Figure 6). Despite the same shape of the QDs, the chevron tails are not sharp but broad [68]. The observed side angle near to 25° corresponds to the (113) crystallographic plane [67, 68].

A cross-sectional TEM image shows that the side of the QD is not a single crystalline plane but has stepped shape (faceting). The steps consist of planes parallel with interface (parallel with {002} crystal planes) and planes with 55° to the interface, corresponding to {111} planes. The envelope curve of the QD cross section is a circle segment with a radius of $R = 64$ nm [68].

The electronic structure of the DE-grown GaAs QDs was investigated by PL spectroscopy [70, 71]. Furthermore, lasing was demonstrated in such structure [72]. After postgrown annealing, improved optical performance was demonstrated [72]. In the case of DE-grown QDs, the better optical properties compared to strain-induced QDs originated from the technology. The DE-grown QDs are prepared at very low substrate temperatures, which are unfavorable in view of the incorporation of undesired background impurities and crystal defects.

4.3. Some aspects to the kinetics of QD growth

The growth kinetic can primarily be investigated by the in situ RHEED tracking. In the annealing phase of QD production, the diffused RHEED pattern becomes spotty nearly simultaneously with the opening of arsenic source. The sufficient arsenic quantity and the low temperature make the build in (infiltration) of As in the Ga phase (i.e., crystallization) possible [73]. This process of infiltration takes about 2–3 min to the sharp chevron image. A crystallized shell comes into being on the Ga droplets. The spotty RHEED pattern originates from the transmission electron diffraction. The electron beam goes through the crystalline GaAs shell layers over the droplet. If there are crystallite formation or droplets on the surface, It is observed that bulk scattering of the grazing beam can occur and the RHEED pattern may become dominated by spots rather than streaks due to transmission electron diffraction (Figure 5) [67].

The scattering from several planes strongly modulates the intensity along the reciprocal lattice rod. Thus, the streaks observed from two-dimensional surface are not observed when transmission dominates [67]. For the transmission case, the reciprocal lattice is an array of points each broadened owing to the finite size of the scattering region. During the annealing, the As diffuses inside of droplets, while excess As builds in (infiltrates) in the shell [73, 74]. Thus, the droplet crystallizes slowly. In the given moment, the rounded shell will be broken by the cornered crystallite grown inside of the structure. At the same time, a chevron-shaped spot develops on the RHEED screen [43, 48, 67, 68].

The angle between two RHEED streaks starting from same reciprocal lattice point is about 55°, as shown in final stage of the QD evolution. These so-called chevron tails are attributed perpendicularly to the facets of the QD [67, 69]. The volume of the QDs is large enough to receive transmission pattern during the electron scattering. The main lateral expansion L_{QD} and height H_{QD} of QD – according to the AFM measurement – are 50 and 5 nm, respectively (see Figure 6). The mean free path of the electrons Λ in GaAs between the crystal planes without collision at the typical incidence angle of RHEED (about 2°) is less than 20 nm. Thus, in our case, there are several (ca. 9) lattice planes to receive transmission character [67].

The analytical TEM investigation provides further contribution to the evolution kinetics. In our TEM investigations, we detected Al content in GaAs QD [48]. Here, this study shows the high-resolution micrograph of a similar QD together with Ga and Al elemental maps of the same area. These images clearly show that the QDs contain both Ga and Al. The presence of Al within the QDs is supported by the explicit protrusion of bright contrast on the Al map at regions corresponding to the QDs. This feature originated from a dissolution process [67]. This process is very important for QR production. This phenomenon will be discussed in the next section.

On the basis of the measurements, the following kinetics explanation seems plausible [68, 69]. The process of GaAs crystallization starts at the edge of the droplet, initialized by the three-phase line (TPL) at this point, serving as discontinuity for the seeding [100]. Although in principle interaction can take place at any point of the droplet, due to the thermal movement, the species, arriving to the edge, will start the seeding of the crystallization process. The described mechanism for this process has been accepted by several authors too. [11, 67]. (Otherwise, it would be difficult to explain the formation of the ringlike QD. In the case of a dot, the seed grows inward, while in the case of ring, it tends to grow outward, which is maintained by the Ga migration.) Since we are dealing with dot shape, the dominant process is arsenic diffusion.

The crystal seed grows inward into the droplet, and also upward simultaneously. This process of growth can only be explained quantitatively because, in the case of nanosizes, the observed bulk processes and properties like diffusion and binding energy cannot be applied. Although similar crystallization processes have been observed, no attempt has been made to explain them until now (see Figure 7) [68]. The growth of the GaAs crystal occurs in direction opposite to the penetration of Al. The crystallization of GaAs can start only after the opening of arsenic cell, while the dissolution of AlAs species occurs immediately upon the deposition of droplets. The process at the surface is quicker than process at the interface. Thus, the process at the surface is the dominant during the processing time of the QD production.

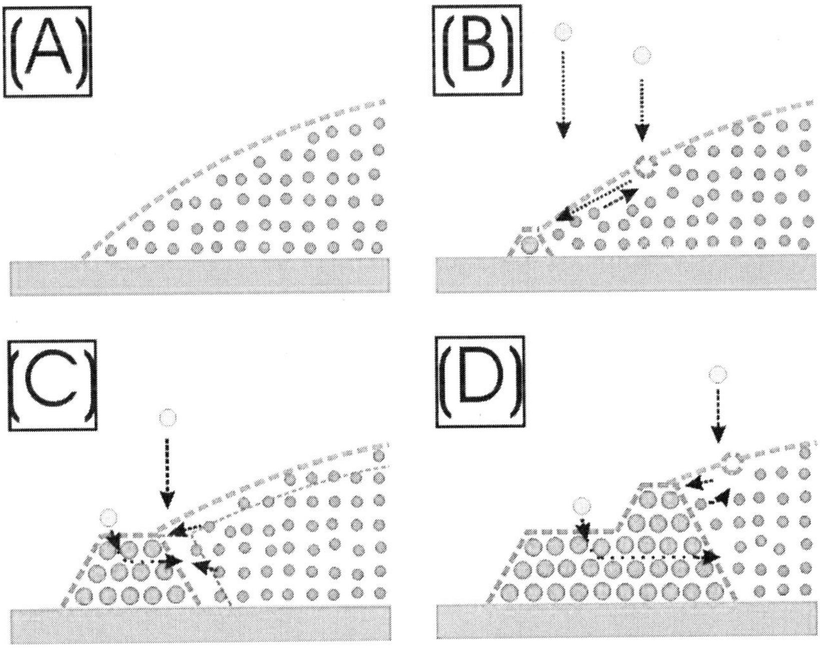

Figure 7. Explanation of the stepped surface evolution.

We explained the evolution kinetics of the stepped outer surface of the QD [68]. The summary of the kinetic is as follows. The intersection of the crystal surface with the droplet is the TPL, which serves as an initial place of crystallization [68]. When a Ga atom of the droplet meets an arsenic atom, they form a GaAs molecule. These GaAs species, making a Brown-like movement over the droplet surface, can reach the TPL, where the crystallization starts (Figure 6). The crystal seed at the TPL will grow on the account of further arriving arsenic species. The outer facet angle of the crystallization center at the edge will be the favorable 55° because it has enough time to find the optimal position (the low index facet). The crystallization seed grows partly upward and also partly in the direction of the droplet inside. During the process of solidification, a circular monocrystalline phase is formed at the droplet edge inheriting the orientation of the perfect substrate. During the process of solidification, the amount of Ga atoms in the droplets decreases so the droplet size decreases too. The outer side of the QD consists of steps of few MLs, where the front panel and the terrace of the step are (111) and (001) planes, respectively.

It is known in fcc crystals that the surface energy of (111) facet is less than that of (001) facet; thus, the latter grows predominantly during crystallization. This takes place by the lateral shift of the low energy (111) step facets with the simultaneous areal growth of (001) facets. With the size decrease of the droplet, its edge moves inward, thus creating a new TPL or crystallization seed at the new place, and the whole process continues as before with the original droplet and substrate. The only difference is that the crystallization takes place along a circle of less and less diameter.

5. RINGLIKE QD PREPARATION

5.1. Growth technology of ringlike QD

The preparation process responsible for the DE formation of ringlike QDs is described in the case of GaAs nanostructures on AlGaAs (001) surface. The growth experiments are performed similarly as described in the previous section, but the technological parameters somewhat differs. On GaAs (001) wafers, first pure GaAs layer is grown, and it is followed by an AlGaAs layer with Al content of 0.3.

Then the GaAs sample with AlGaAs layer must be cooled down to 300°C. On the surface, Ga is deposited as described in former section. Then θ = 3.75 ML gallium is deposited with the flux of 0.19 ML/s without any arsenic flux. During the annealing, the temperature remained the same (300 ºC), but the As pressure was 4 × 10^{-6} Torr. The production of the quantum objects was tracked continuously in the direction of [11 ¯ 1¯ 0] with the help of RHEED [67].

Further, two types of ringlike nanostructures were generated with different amounts of deposited Ga. One of them was generated at 570°C on AlGaAs (001) surface applying 6.4 ML Ga. In this case, the

AFM measurement shows ringlike QD with deep hole in the middle of the structure and very large clusters [95]. The other type of the nanostructures was prepared similarly, but the deposited Ga was less than that in the former case. The Ga coverage was 3.2 ML. In this case, the AFM picture shows special-shaped nanostructure with very deep crater in the middle surrounded by ringlike bulge formations and also sallow nanocraters, with plane rims (without bulge) [75, 76].

After the growth process, quantum objects are investigated using the AFM method, and the first mentioned types of nanostructure are studied with PL technique, too. Temperature-dependent PL spectra measured on the GaAs ringlike QD samples are compared with the conventionally shaped QDs.

The technology can continuously be tracked by RHEED. The initial stage of the surfaces is the same as in the case of conventional QD. The RHEED pattern of the initial surface shows sharp streaks (stage t_0 on Figure 8). After the Ga deposition, the pattern diffuses on the RHEED screen similarly as in the case of QDs (stage t_1). The deposited Ga is in liquid phase. The disappearance of the RHEED pattern originates from the appearance of the liquid phase of Ga droplet on the surface. However, after the Ga deposition, the change of the observed RHEED pattern is quite different. After the offering of arsenic background of 4×10^{-6} Torr, the RHEED pattern develops very slowly, over 5 min. Contrary to the case of conventionally shaped QDs, which is almost at the same time of the arsenic cell opening, the RHEED pattern changed suddenly. The developed pattern contains in the middle a streak with a small spot and around elongated larger spots. According the AFM measurement, the density of the ringlike structures is 1.5×10^9 cm^{-2}. In this case, the effect of open surface on RHEED is larger than in the case of QD [69]. It is shown that the characteristic RHEED pattern of QD is still recognizable even if QD density is one order of magnitude less [43].

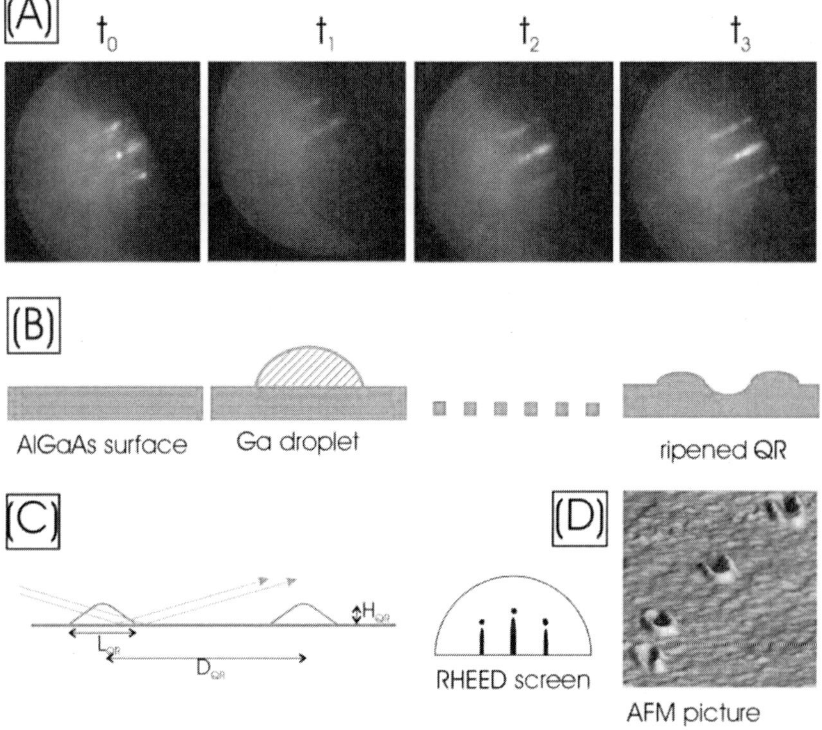

Figure 8. (A) During the evolution of ringlike DE-grown QD, the changing of the RHEED pattern and (B) its temporal evaluation. (C) Interaction of ringlike QD and electron beam. (D) AFM picture of DE-grown ripened nanostructures.

5.2. Geometry and electronic structure of the ringlike QD

The shapes of ringlike QDs are various. The main parameters are the following: ring-middle diameter, width of the ring base, and height. As we will show, the facet angle is determined by the volume. The further shape parameter is the level of the ring middle. The ring middle can lie higher or deeper than the original substrate surface. At lower annealing temperature and at less arsenic ambient pressure, the droplet metamorphosis results in ringlike nanostructure with smaller crater in the middle [78–81]. For example, at 4.75 ML Ga deposition, under 1×10^{-5} Torr arsenic pressure and 350°C annealing for 10 min, the resulted structures are ring with shallow crater [79].

In another case, if the ring center lies deeper [75, 77], the structure is called a nanohole. In this case, thermal etching is dominant. If the annealing temperature is higher and the arsenic ambient pressure is smaller, then we receive ring structure with low-lying center. For example, at 8 ML deposited Ga, at 520°C annealing and under various arsenic pressure, we can get nanostructures with various deepness in the middle. If the arsenic pressures are 3×10^{-6}, 2×10^{-6}, 1×10^{-6}, and 1×10^{-7} Torr, then the holes are ca. 3, ca. 7, ca. 8.5, and 9.5 nm deep, respectively [62, 75, 76, 82].

The electronic structure is strongly governed by the shape of the nanostructure. Here, the advantages of the ringlike QD compared to the conventionally shaped QD are described using temperature-dependent PL spectra. Under measurement, the temperature ranges between 4 K and 300 K. The samples are excited by Ar^{+} ion laser at 488 nm wavelength. The resolution was better than 0.5 nm [83].

The PL spectra of the uncovered GaAs ringlike QD grown on AlGaAs surface are shown in Figure 9. At 4 K, the spectrum has five peaks. They can be explained as bound exciton (1.5129 eV), exciton bound to acceptor (1.4892 eV), and its longitudinal optical phonon replica (1.4577 eV). Further on at higher energies, two peaks appear (1.5308 and 1.5602 eV). The PL spectra were also recorded as a function of temperature. To verify the identification of the peaks, the temperature dependence of the band-gap energy of GaAs and that of the bound exciton were compared. At 4 K in the case of conventionally shaped QD with identical volume, three peaks can only be seen at lower energy range. It is visible that the quantum confinement in ringlike QD is larger than that in QD at same nanostructure volume. In the following, this phenomenon will be explained [83].

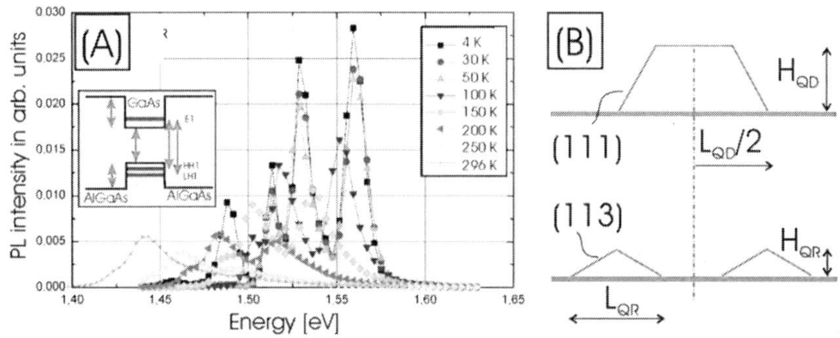

Figure 9. Temperature-dependent PL spectra of ringlike QD. (B) The explanation, why the ringlike QD has significantly sharper PL peaks compared to conventionally shaped QD.

The base diameter of the QD (L_{QD}) investigated is much larger, and its height (H_{QD}) is also larger but comparable. Thus, here no quantum effects are expected. Therefore, it was impossible to detect a peak shift or peak broadening for QD. At the same volume, for ringlike structure, the laterally wide (L_{QR}) nanostructure had a height (H_{QR}) significantly less than 7 nm. Consequently, the supposition of quantum confinement produces proper result [83].

The PL spectra show characteristic peak width, depending on the temperature and the size distribution of these structures. At the same temperature, a broader size variation results in a broader PL peak. Under similar growth conditions, the PL peaks of ringlike structures have significantly narrower full width at half maximum (FWHM) (less than one third) than that of the conventionally shaped QDs [78, 84].

The sharper PL peaks can be explained by follows. As it is known, both kinds of QDs is formed from a gallium droplet. Assume that the volume and its variation of the initial droplets are the same in both cases. The facet of the nanostructures is size dependent and cannot be arbitrarily sized [43]. For the QD, let's start from the greater (111) facet (the approx. diameter ($2r = L_{QD}$) is 100 nm [43]). For the ringlike QD, only the (113) facet can be taken into account because of the smaller volume (the approximate width (L_{QR}) is 60 nm [43, 44, 83], dedicated to a circle segment). Due to the crystallographic constraints, the geometry of the formed QDs is determined. The height-to-diameter ratio cannot be arbitrary; it can be defined by a single parameter.

If r ($r = L_{QD}/2$) is the radius of the base circle belonging to the initial droplet, the volume of the developing QD, as a function of r, is given by $V = 1.58r^3$. For the ringlike nanostructure, from the equality of the volumes, the w ($w = L_{QR}$) parameter can be calculated as $w = 0.71r$. (This means that the nano-object with a parameter of $2r = 100$ nm has (111) facets, and the other one with a parameter of $w = 0.71 \times 100 \sim 70$ nm or less has (113) facets. It corresponds to the above presented measured data.) It means, that the height of the nanostructure influences commonly the quantum behavior. The heights of the QD and QR structures can be expressed as a function of their volumes: $m_{QD} = V^{1/3}/1.34$ and $m_{QR} = V^{1/3}/6.56$, respectively. Assuming the same variations of the volumes, it can be seen that the variation of the height for QR is much smaller than for QD; consequently, the corresponding FWHM of the PL peak is smaller [83].

5.3. Some aspects to the formation of ringlike nanostructure

The primer information to discover the formation kinetic originates from RHEED tracking [67]. After the deposition of Ga on AlGaAs (001) surface, the RHEED picture is becoming diffused due to the amorphous nature of the phase present on the surface. The annealing phase begins after the offering of arsenic component ($p_{As} = 4 \times 10^{-6}$ Torr, $T_{sub} = 300°C$). After releasing the arsenic, some time is needed for the development of the characteristic sharp pattern, representative of the crystalline structure. This is an indication that the liquid state on the surface stays longer and that the material transfer processes helps the formation of the ringlike structures. For detailed technological parameters of the different DE nanostructures, we refer the readers to the literature [32–35, 43–45, 62].

After the completion of the growth, the QR structures were investigated with AFM. The perspective AFM image and the top view with line scans are shown in Figure 10. The dimensions of QRs were determined from individual line scans. The density of the nanostructures determined from AFM pictures was 1.5×10^9 cm^{-2}. It is visible that the middles of the nanostructures are deeper than the original surface level [85]. It can be shown, from the AFM measurement that although the shapes and sizes of QRs are fairly uniform, we can observe small deviations from these averages. It is often observed that the smaller diameter rings have deeper depressions in the center and the larger diameter rings have shallower ones in the middle. (In the illustrations, the smaller and larger objects are labeled with "S" and "L", respectively.) The QRs have slightly elongated shapes due to the different binding properties in [110] and [11 ¯ 1¯ 0] directions [86].

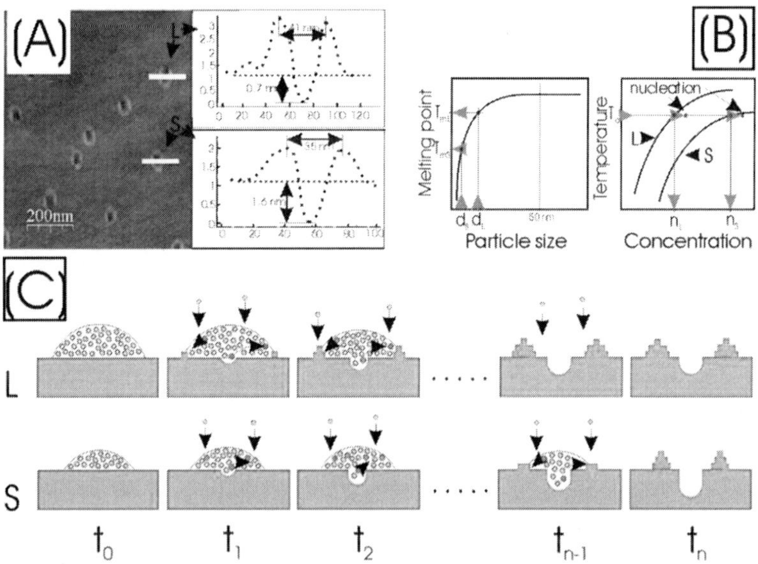

Figure 10. (A) Cavity dependence on the droplet size. (B) Functions of melting point and solubility. (C) Temporal evaluation of the smaller and larger ringlike QDs.

These facts are contradictory to the aspect ratios of ringlike QDs, and the explanation can help to understand the kinetics of the DE quantum structure formation. The explanation to this extraordinary behavior is as follows [85]. It is an obvious assumption that the larger droplets are leading to the development of rings of larger diameter and equally the smaller droplets to smaller rings. The intersection of the crystal surface with the droplet edge is the so-called TPL, which is the initial place of crystallization [85]. The TPL of larger diameter initiates a larger droplet, and equally, the smaller diameter forms from a smaller droplet. It is known from the liquid phase epitaxy that thermal etching takes place at the interface of the Ga melt and AlGaAs surface. This phenomenon was confirmed by analytical TEM for DE QD [69]. The Ga melt can solve the arsenide molecules (e.g., GaAs). These arsenide molecules originate partly from the thermal etching of the AlGaAs substrate surface and partly from the reaction of the external (from effusion source) arsenic atoms. (When a Ga atom of the droplet meets an arsenic atom from the environment, they form a GaAs molecule.) These molecules, due to thermal movement in the droplet, can reach the TPL, where the crystallization takes place. During the process of solidification, the material migrates from the middle to the edge of the nanostructure. A circular crystalline phase is formed at the droplet edge. The proposed kinetics for the formation is shown schematically in Figure 10.

It is known that the melting point decreases with the reduction of the particle size. The normalized melting curve versus the diameter of the particle (Figure 10) shows that when the particle size is less than 50 nm, then the melting point depends very strongly on size. In the nanorange, this dependence on the size is stronger than for the bulk. What makes it more complicated is the fact that the melting point depends on the particle shape as well. In this case, the Ga droplet's shape is a segment of a sphere. Its width in the middle falls in the range of 10–20 nm. Here, the change in the melting point is particularly sharp [85].

Around the edges the structure is thin. The experimentally obtained melting curves for near spherical metal nanoparticles show similarly. We use these curves for the qualitative assessment. This indicates that the melting point of the large and that of the small Ga droplets can differ considerably. The solubility curves for different particle sizes (Figure 10) show that at the same temperature, the larger droplet has lower saturating concentration than the smaller one. The meaning of this is that crystallization in the larger droplet will take place earlier, at lower arsenide concentration than in a smaller one. The smaller droplet will crystallize later, during arsenide concentration.

The temporal evaluation of the smaller and larger ringlike QDs is shown in lower part of the figure onset (Figure 10) [85]. In other words, in the larger droplets, the probability of the formation of the crystallization seeds is higher; therefore, the crystallization takes place earlier so less time is spent on material transportation, causing the development of the depression in the middle. When the droplets are small, the probability is less, and crystallization starts later, leaving more time for the formation of deeper depression in the center of the ring. This process is influenced by other factors as well. The melting temperature of the nanostructure is dropping with its diminishing size, staying longer in liquid state at the same temperature, spending more time on the formation of deeper in the center [85].

The above-described finding can be proven by further experiments (in Section 5.1, ringlike nanostructure production is described where the amounts of the deposited Ga were 3.2 and 6.4 ML, respectively). The explanation is as follows [85, 87, 88]: At a given temperature, there is a critical droplet size (CDS) under which the solution begins. After the Ga deposition, droplets form, followed by the growth of the larger droplets at the expense of smaller ones according the Ostwald ripening. When the droplet is smaller than the CDS, the substrate solution begins. In the case of Ga 6.4 ML, the formation of small holes and large clusters can be observed. The sizes of the droplets formed are above the CDS. After the deposition, the differentiation of the droplets begins. The smaller droplets reach critical size and start solving the substrate. This state is frozen via opening the arsenic cell.

When the deposited Ga is 3.2 ML, we can observe shallow holes with plane rims. The explanation is as follows [85, 87, 88]. In this case, the quantity of the deposited Ga is small. The formed droplets are under CDS. Therefore, the solution starts under the droplets immediately after the droplet formation. Under the small droplets, the solution is faster, but the material of liquid Ga is used up in a short time. The reduction in material is due to, first, the solution and, second, the migration of materials toward the larger droplets. After a short time, at the smaller droplets, the solution stops, while it carriers on further under the larger ones. The larger droplets will not be spent, and therefore the surrounding ring will freeze after the opening of the arsenic cell.

6. SPECULAR-SHAPED QDS AND COMPLEXES

6.1. The inverted QDs

The ringlike QD is an unconventional QD, which has advantages over the conventional QD. As we have shown, the DE technique is a many-sided tool in this field. It allows us to fabricate further interesting nanostructures. Here, we will further discuss unconventional-shaped QDs and their complexes. In the present subsection, we will focus on an alternative QD preparation technique. This is not only an alternative preparation but the QD with inverted technology can also have advantages at special application.

A further recent method for the fabrication of strain-free QD is the nanohole filling. The nanohole is created by localized thermal etching, and it is filled subsequently [95]. This is a QD with inverted technology. The localized thermal etching takes place at conventional MBE growth temperatures, and we expect only very low level of crystal defects. The nanoholes are created in a self-organized fashion by local material removal [89]. For the inverted QD fabrication, nanoholes are generated using Al droplets on AlAs surface. Subsequently, the holes are filled with GaAs layer to form QDs of controlled height [90]. The nanoholes are filled with GaAs in pulsed mode.

Here, we show a cross-sectional study of an inverted QD [66]. The technology of this QD is as follows. The structure is grown on GaAs (001) surface. On the surface, AlGaAs and AlAs layer sequence is grown. The AlGaAs layer has 0.23 Al content. The AlAs and the AlGaAs layers are 5.5 and 19.5 nm, respectively. After the growth of every AlAs layer, 3.2 ML Al is deposited at 650°C without any arsenic flux. The deposited Al MLs form droplets on the surface. This is followed by an annealing step of 180 s, during which the droplets transform into nanoholes.

Following that, the holes are partially filled by the deposition of 0.6 nm thick GaAs in a growth-interrupted fashion. During the hole filling, the substrate temperature is 600°C. Depending on the foreseen high of the QD, some growth of 0.5 s and pause of 30 s sequences are carried out. Finally, the GaAs QD can be covered with AlGaAs. The scheme and the TEM image of the inverted QD structure are shown in Figure 11. The above-described pulsed technique allows us to create not only single QDs but also vertically aligned QD pairs [91].

The distances on the TEM image correspond with the intended layer thicknesses. The picture shows GaAs QD embedded in AlGaAs barrier material. Furthermore, the TEM image shows a thickening of the AlAs layer in the surrounding the nanohole. Elementary maps taken using electron energy loss spectroscopy confirm that the thickened region consists of AlAs. We identify the additional AlAs on top of the flat AlAs layer as the wall that surrounds the nanohole opening. Furthermore, the TEM image on Figure 11 shows that the next AlAs layer grown on top of the inverted QD layer is bended upward at the location of the QD, resulting in the formation of a kind of hill [66].

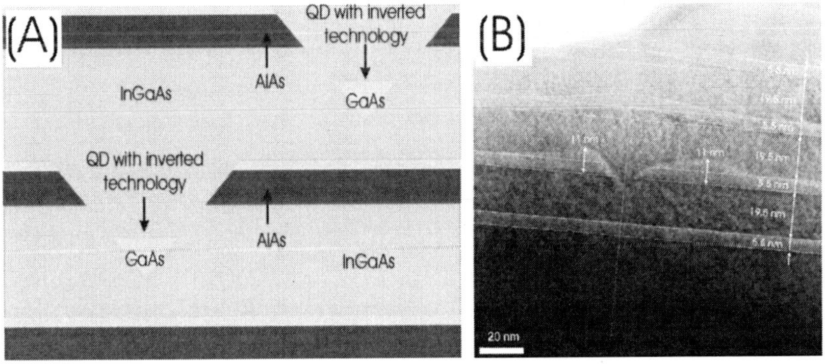

Figure 11. (A) Scheme of the sample with inverted QDs. (B) Cross-sectional TEM image of a GaAs QD embedded in InGaAs layer.

6.2. The laterally aligned QD pairs

Two semiconductor QDs in close proximity, which can each spatially confine an individual charge carrier in a discrete energy level, interact quantum mechanically with each other. In particular, the wave functions of the charge carriers confined in each QDs of the pair begin to overlap, resulting in an efficient tunneling. Furthermore, the wave functions may become mixed to develop molecular orbital. Moreover, resonance in the optical range leads to the formation coupled QD pairs with the help of dipole–dipole interaction. These research leads toward quantum information processing. The QD pairs and their systems offer, at least conceptually, the potential of implementing scalable arrays of qubits.

In this part, we are dealing GaAs QD pairs prepared on AlGaAs surface with the usage of the anisotropy of the (001) oriented surface. We will show two preparational series. One of them is carried out under lower temperature, at less amount of deposited MLs. The other ones is prepared under higher temperature at higher amount of deposited Ga.

In the first case, AlGaAs with Al content of 0.27 is grown on the GaAs (001) surface. After this, Ga droplets are created at 330°C temperature on the substrate. The crystallization happens at 200°C,

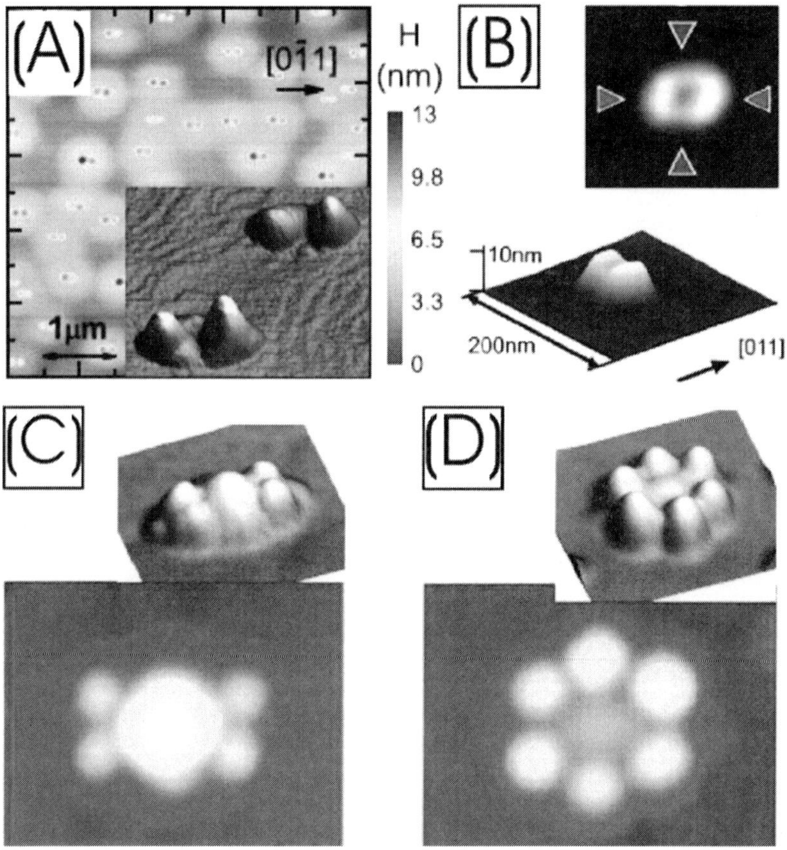

Figure 12. QD pairs grown at high substrate temperature with large amounts of Ga MLs. (Illustration originated from R. Pomraenke et al.; Phys. Rev. B 77 (2008) 075314) [105].) (B) (B) QD pairs grown at low substrate temperature with low amounts of Ga MLs. (Illustration originated from M. Yamagiwa et al.; Appl. Phys. Lett. 89 (2006) 113115) [104].) (C and D) Growth of quad- and hexa-QD molecules initiated by GaAs mound. (Illustrations originated from J.H. Lee et al.; Appl. Phys. Lett. 89 (2006) 202101) [110].)

under accurate control of the arsenic flux [104]. The ripened structure basically consists of two QDs aligned in the [0 1⁻ 1⁻ 1] crystalographic direction. The average base size and height of each

QD are 45 and 10 nm, respectively. The QDs are separated by an average distance of 39 nm between their apexes, as measured by AFM. The density of the structure is 2×10^8 cm^{-2} [104]. For the study of its optical properties, the QD pairs are embedded in AlGaAs barrier layer.

The second technology is carried out on GaAs surface. First, AlGaAs layer is grown with 0.3 Al amount. At 550°C substrate temperature, a large amount of Ga is deposited to create droplets on the surface. The structure is "arsenized" by fine control of the flux [99]. The resulting dots are rather large. The individual pairs have an interdot distance of about 130 nm and are aligned along the $[0\ \bar{1}\ \bar{1}\ 1]$ direction. The dots with a height of 10 nm reside on a shallow base with a diameter of 300 nm. The density of the structure is 2.3×10^8 cm^{-2} [99]. For optical characterization, the QD pairs are buried by the deposition of AlGaAs layer.

In the first case, the micro-PL spectra of a single QD pair show the ensemble of emissions, which may indicate the existence of a tunnel coupling between the members of the pair. However, it must be mentioned that these emission lines are still not fully understood [104]. At the second experiment, the PL study is performed at a rather larger inter-dot separation. Here the formation of coherently coupled molecular QD states is suppressed, and excitonic interactions between neighboring QDs within each members are weak [105]. The process of the QD pair evolution is also not fully understood. Instead of the hemispherical shape of the initial droplet, the ripened structure is anisotropic. The explanation is based on the anisotropic surface potential of the GaAs (001) surface (Figure 12).

6.3. Configuration of QD ensembles

Until now, the most widely studied system of interacting self-assembled QDs is a vertically stacked QD molecule. However, this system is limited by its one-dimensional nature and is not easily scalable. Alternatively, a variety of techniques can be used to modify the substrate to nucleate laterally coupled QD molecules in two dimensions by MBE technology. Various shape droplets can subsequently serve as templates for further growth, without the need for further surface preparation.

The DE-like created nanohole and also the homo-DE-grown island can serve as an initialization place for the growth of further nanostructures. The nanohole as a template for QD molecules is demonstrated by Songmuang and coworkers [106]. The nanohole can serve as a template for QD pair too [107]. Salamo and coworkers proposed a new way for the preparation of QD molecules [92-94, 110]. They adopt a hybrid growth approach, utilizing both DE and strain-induced growth to overcome some limitations of the Stranski–Krastanov QD growth mode alone on a planar GaAs surface [102]. Using MBE, a self-assembled InGaAs QD molecule is realized around GaAs mound formed by DE on GaAs (001) surface [102].

The number of QDs per GaAs mound can be effectively controlled by varying the InAs ML coverage. The number of QDs per template ranges from two to six. The technology is as follows. On GaAs (001) surface, 3 ML Ga is deposited at 500°C without any arsenic flux to form Ga droplets. Subsequently, 80 s of annealing occurs, and the substrate temperature is decreased to 150°C. The Ga droplet is crystallized under 1.3×10^{-5} Torr equivalent pressure for 100 s. During this time, nanoscale GaAs mound forms as template. The substrate temperature is raised again to 500°C, and InAs deposition follows. If the deposited InAs are 1.4, 1.6, and 2.0 ML, then the created nanostructures are bi-QD molecule, quad-QD molecule, and hexa-QD molecule [108].

During the QD molecule preparation, the height of the GaAs mound decreases from the original value, and also the diameter decreases. It appears that the InAs growth started mixing with the Ga atoms from the GaAs mounds, resulting in InGaAs shoulders on initial templates. This technique allows to fabricate the so-called quantum-rod pair and lip-shaped structures too [108]. The technology is extended with incident angle-controlled molecular beam technique [109].

7. ACKNOWLEDGEMENTS

In alphabetical order, the author would like to thank J. Balázs, J. Bozsik, A. Csík, M. Csutorás, L. Dobos, Z. Hajnal, Ch. Heyn, W. Hansen, P. Kucsera, J. Makai, C. Manzoni, B. Pécz, B. Pődör, I. Réti, K. Sakoda, S. Sanguinetti, A. Stemmann, J. Takács, G. Tényi, L. Tóth, and A. Ürmös for their fruitful cooperation and to the "Concert-Japan ERA-NET FemtoTera" Project for financial support via OTKA-NN 114457 Found.

REFERENCES

1. D. Leonard, M. Krisnamurthy, C. M. Reaves, S. P. Denbaas, P. M. Petroff; *Appl. Phys. Lett.* 63 (1993) 3203.
2. V. Bressler-Hill, S. Varma, A. Lorke, B. Z. Nosho, P. M. Petroff, W. H. Weinberg; *Phys. Rev. Lett.* 74 (1995) 3209.
3. W. Yang, H. Lee, T. J. Johnson, P. C. Sercel, A. G. Norman; *Phys. Rev. B* 61 (2000) 2784.
4. O. Stier, M. Grundmann, D. Bimberg; *Phys. Rev. B* 59 (1999) 5688.
5. Ch Heyn, A. Bolz, T. Maltezopoulos, R. L. Johnson, W. Hansen; *J. Cryst. Growth* 278 (2005) 46.
6. N. Koguchi, S. Takahashi, T. Chikyow; *J. Cryst. Growth* 111 (1991) 688.
7. T. Mano, K. Watanabe, S. Tsukamoto, H. Fujikoa, M. Oshima, N. Koguchi; *Jpn. J. Appl. Phys.* 38 (1999) L1009.
8. J M Lee, D H Kim, H Kong, J C Woo, S J Park; *J. Cryst. Growth* 212 (2000) 67.
9. T Mano, T Kuroda, K Mitsuishi, M Yamagiwa X J Guo, K Furuya, K Sakoda, N Koguchi; *J. Cryst. Growth* 301–302 (2007) 740.

10. T. Mano, S. Tsukamoto, H. Fujioka, M. Oshima, N. Koguchi; *J. Cryst. Growth* 227–228 (2001) 1069.

11. Z. Gong, Z. C. Nin, S. S. Huang, Z. D. Fang, B. Q. Sun, J. B. Xia; *Appl. Phys. Lett.* 87 (2005) 093116.

12. T. Kuroda, T. Mano, T. Ochiai, S. Sanguinetti, K. Sakoad, G. Kigo, N. Koguchi; *Phys. Rev. Lett.* 72 (2005) 205301.

13. S. Hwang, Z. Nin, Z. Fang, H. Ni, Z. Gong, J. B. Xia; *Appl. Phys. Lett.* 89 (2006) 031921.

14. M. Yamagiwa, T. Mano, T. Kuroda, T. Tateno, K. Sakoda, G. Kido, N. Koguchi; *Appl. Phys. Lett.* 89 (2006) 113115.

15. J. H. Lee, Zh. Wang, N. W. Strom, Yu. I. Mazur, G. J. Salamo; *Appl. Phys. Lett.* 89 (2006) 202101.

16. H. Lüth: Surfaces and Interfaces of Solid Materils; Springer Verlag, Berlin (1998).

17. J. Tersoff, F. K. LeGoues; *Phys. Rev. Lett.* 72 (1994) 3570.

18. J. M. Gérard, O. Cabrol, J. Y. Marzin, N. Lebouché, J. M. Moison; *Mat. Sci. Eng. B* 37 (1996) 8.

19. D. Bimberg, M. Grundmann, N. N. Ledentsov: Quantum Dot Heterstructures, John Wiley, New York (1999).

20. B. A. Joyce, D. D. Vvedensky: Growth modes and morphologies on GaAs surfaces, (in Thin Films: Heteroepitaxial Systems, Eds.: W. K. Lin, M. B. Santos) World Sci., Singapore (1999) p 368.

21. V. A. Shcukin, D. Bimberg; *Rev. Mod. Phys.* 71 (1999) 1125.

22. P.Ribas, V. Krishnamoorthy, R.M. Park; *Appl. Phys. Lett.* 57 (1990) 1040.

23. E.A. Fitzgerald; *Mater. Sci. Rep.* 7 (1991) 621.

24. Á. Nemcsics, F. Riesz, L. Dobos; *Thin Solid Films* 343 (1989) 520.

25. Á. Nemcsics, F. Riesz; *Phys. Stat. Sol. C* 0 (2003) 893.

26. R. Schnurpfeil, K. Restöft, A. Müller, Á. Nemcsics, R. Manzke, M. Skibowski; *J. Electr. Spectr. Rel. Phen.* 68 (1994) 175.

27. Á. Nemcsics, J. Olde, M. Geyer, R. Schnurfpeil, R. Manzke, M. Skibowski; *Phys. Stat. Sol. A* 155 (1996) 427.

28. S. Lee, I. Daruka, C.S. Kim, A-L Barabási, J.L. Metz, J.K. Furdyna; *Phys. Rev. Lett.* 81 (1998) 3479.

29. N. Koguchi, S. Takahashi, T. Chikyow; *J. Cryst. Growth* 111 (1991) 688.

30. N. Koguchi, K. Ishige; *Jpn. J. Appl. Phys.* 32 (1993) 2052.

31. T. Mano, T. Kuroda, K. Mitsuishi, M. Yamagiwa, X. J. Guo, K. Furuya, K. Sakoda, N. Koguchi; *J. Cryst. Growth* 301–302 (2007) 740.

32. T. Mano, S. Tsukamoto, H. Fujioka, M. Oshima, N. Koguchi; *J. Cryst. Growth* 227–228 (2001) 1069.

33. Z. Gong, Z. C. Nin, S. S. Huang, Z. D. Fang, B. Q. Sun, J. B. Xia; *Appl. Phys. Lett.* 87 (2005) 093116.

34. T. Kuroda, T. Mano, T. Ochiai, S. Sanguinetti, K. Sakoad, G. Kigo, N. Koguchi; *Phys. Rev. Lett.* 72 (2005) 205301.

35. S. Hwang, Z. Nin, Z. Fang, H. Ni, Z. Gong, J. B. Xia; *Appl. Phys. Lett.* 89 (2006) 031921.

36. W.Z. Ostwald; *Z. Phys. Chem.* 34 (1900) 975.

37. M. Zinke-Allmang, L.C. Feldman, S. Nakahara; *Appl. Phys. Lett.* 51 (1987) 975.

38. M. Zinke-Allmang, L.C. Feldman, W. van Saarloos; *Phys. Rev. Lett.* 68 (1992) 2358.

39. G.Z. Pan, K.N. Tu; *Appl. Phys. Lett.* 68 (1996) 1654.

40. G.R. Carlow, M. Zinke-Allmang; *Phys. Rev. Lett.* 78 (1997) 4601.

41. K. Shorlin, S. Krylov, M. Zinke-Allmang; *Physica A* 261 (1998) 248.

42. A. Raab, G. Springholz; *Appl. Phys. Lett.* 77 (2000) 2991.

43. Ch. Heyn, A. Stemmann, A. Schramm, H. Welsch, W. Hansen, Á. Nemcsics; *Appl. Phys. Lett.* 90 (2007) 203105.

44. Ch. Heyn, A. Stemmann, A. Schramm, H. Welsch, W. Hansen, Á. Nemcsics; *Phys. Rev. B* 76 (2007) 075317.
45. J.H. Lee, Z.M. Wang, N.W. Strom, Y.I. Mazur, G.J. Salamo; *Appl. Phys. Lett.* 89 (2006) 202101.
46. A. Ohtake; *Surf. Sci. Rep.* 63 (2008) 295.
47. Q.K. Xue, T. Hashizume, T. Sakurai; *Prog. Surf. Sci.* 56 (1997) 1.
48. D.K. Biegelen, R.D. Brinans, J.E. Northrup, L.E. Swartz; *Phys. Rev. B* 41 (1990) 5701.
49. Á. Nemcsics, J. Takacs; *Acta Polytechn. Hung.* 7 (2010) 109.
50. Á. Nemcsics, J Takacs; *Semiconductors* 45 (2011) 91.
51. M. Zinke-Allang, L.C. Feldman, M.H. Grabow; *Surf. Sci. Rep.* 16 (1992) 377.
52. V. A. Shcukin, D. Bimberg; *Rev. Mod. Phys.* 71 (1999) 1125.
53. M. Wautelet, D. B. Beljonne, J.-L. Brédas, J. Cornil, R. Lazzaroni, P. Lecére, M. Alexander, P. Dubois, P. Gills, Y. Gossuin, R. Muller, A. Ouakssim, A. Roch, D. Duviver, J. Robert, R. Gouttebaron, M. Hecq, F. Monteverde: Nanotechnologie; Oldenbourg Verlag, München (2003).
54. I. Grossy, M.B. Small; *J. Cryst. Growth* 11 (1971) 157.
55. H. Ijmin, S. Gonda; *J. Cryst. Growth* 33 (1976) 215.
56. N.S. Peev; *J. Cryst. Growth* 98 (1989) 499.
57. E. Lendvay, T. Görög, V. Rakovics; *J. Cryst. Growth* 72 (1985) 616.
58. Á. Nemcsics, L. Tóth, L. Dobos, Ch. Heyn, A. Stemmann, A. Schramm, W. Hansen; *Superlatt. Microstr.* 48 (2010) 351.
59. Zh. M Wang, B.L. Liang, K A Sablon, G.J. Salamo; *Appl. Phys. Lett.* 90 (2007) 113120.
60. Ch. Heyn, A. Stemmann, W Hansen; *Appl. Phys. Lett.* 95 (2009) 1731100.

61. Ch. Heyn, A. Stemmann T Köppen Ch Strelow, T Kipp, M Grave, S Mendach, W Hansen; *Appl. Phys. Lett.* 94 (2099) 183113.

62. Ch Heyn, A Stemmann, R Eiselt, W Hansen; J. Appl. Phys. 105 (2009) 054316.

63. A. Stemmann, Ch. Heyn, T. Köppen, T. Kipp, W. Hansen; *Appl. Phys. Lett.* 93 (2008) 123108.

64. Ch. Heyn, A. Stemmann, W. Hansen; *J. Cryst. Growth* 311 (2009) 1839.

65. Ch Heyn, A. Stemmann, T. Köppen, Ch. Strelow, T. Kipp, M. Grave, S. Mendach, W. Hansen; *Nanoscale Res. Lett.* 5 (2010) 576.

66. Á. Nemcsics, Ch. Heyn, L. Tóth, L. Dobos, A. Stemmann, W. Hansen; *J. Cryst. Growth* 335 (2011) 58.

67. Á. Nemcsics, Ch. Heyn, A. Stemmann, A. Schramm, H. Welsch, W. Hansen; *Mat. Sci. Eng. B* 165 (2009) 118.

68. Á. Nemcsics, L. Tóth, L. Dobos, A. Stemmann; *Microel. Reliab.* 51 (2011) 927.

69. D. Choi, M. Warusawithana, C.O. Chui, J. Chen, W. Tsai, D.G. Schlom, J.S. Harris; *Mat. Res. Symp. Proc* 996 (2007) H5–H31.

70. H. Lee, R. Lowe-Webb, W. Yang, P.C. Sercel; Appl. Phys. Lett. 72 (1998) 812.

71. K. Watanabe, N. Koguchi, Y. Gotoh; Jpn J Appl. Phys 39 (2000) L79.

72. T. Mano, T. Kuroda, M. Yamagiwa, G. Kido, K. Sakoda, N. Koguchi; *Appl. Phys. Lett.* 89 (2006) 183102.

73. Á. Nemcsics; *Semiconductor* 39 (2005) 1352.

74. I. Mojzes, T. Sebestyén, P. B. Barna, G. Gergely, D. Szigethy; *Thin Solid Films* 61 (1979) 27.

75. N. Koguchi, S. Takahashi, T. Chikyow; *J. Cryst. Growth* 111 (1991) 688.

76. N. Koguchi, K. Ishige; *Jpn. J. Appl. Phys.* 32 (1993) 2052.

77. J.M. Garcia, D. Grandos, J.P. Silveira, F Briones; *Microel. Journal* 35 (2004) 7.

78. T. Mano, T. Kuroda, M. Yamagiwa, G. Kido, K. Sakoda, N. Koguchi; *Appl. Phys. Lett.* 89 (2006) 183102.

79. C. Zhao, Y.H. Chen, B. Xu, P. Jin, Z.G. Wang; *Appl. Phys. Lett.* 91 (2007) 033112.

80. S. Sanguinetti, T. Mano, A. Gerosa, C. Somaschini, S. Bietti, N. Koguchi, E. Grilli, M. Guzzi, M. Gurioli, M. Abbarachi; *J. Appl. Phys.* 104 (2008) 113519.

81. C. Somaschi, S. Bietti, N. Koguchi, S. Sanguinetti; *Appl. Phys. Lett.* 97 (2010) 203109.

82. A Stemmann, T Köppen, M Grave, S Wildfang, S Mendach, W Hansen, Ch Heyn; *J. Appl. Phys.* 106 (2009) 064315.

83. Á. Nemcsics, J. Balázs, B. Pődör, J. Makai, A. Szemann; Phys. Stat. Sol. C 8 (2011) 2826.

84. K. Watanabe, S. Tsukamoto, Y. Gotoh, N. Koguchi; *J. Cryst. Growth* 227 (2001) 1073.

85. Á. Nemcsics, A. Stemmann, J. Takács; *Microel. Reliab.* 52 (2012) 430.

86. Á. Nemcsics; *J. Cryst. Growth* 217 (2000) 223.

87. Á. Nemcsics; *AIP Conf. Proc.* 1598 (2014) 79.

88. Á. Nemcsics; *Acta Politechn. Hung.* 8 (2011) 5.

89. Ch. Heyn, A. Stemmann, T. Köppen, Ch. Strelow, T. Kipp, M. Grave, S. Mendach, W. Hansen; *Appl. Phys. Lett.* 94 (2009) 183113.

90. Z.M. Wang, B.L. Liang, K.A. Sablon, G.J. Salamo; *Appl. Phys. Lett.* 90 (2007) 113120.

91. V. Pollojärvi, A. Stemmann, M. Guina, A. Stemmann, Ch. Heyn; *Nanotechnology* 22 (2011) 105603.

92. J.H. Lee, S. Sablon, Z.M. Wang, G.J. Salamo; *J. Appl. Phys.* 103 (2008) 054301.

93. M. Hanke, M. Dubslaff, M. Schmidbauer, Z.M. Wang, Y.I. Mazur, P.M. Lytvyn; *Appl. Phys. Lett.* 95 (2009) 023103.

94. M. Dubslaff, M. Hanke, M. Burghammer, S. Schröder, R. Hoppe, C.G. Schroeder, Y.I. Mazur, Z.M. Wang, J.H. Lee, G.J. Salamo; *Appl. Phys. Lett.* 98 (2011) 213105.

95. Ch. Heyn; *Phys. Rev. B* 83 (2011) 165302.

96. R. Viswanatha, S. Saha-Dasgupta, D.D. Sarma; *Phys. Rev. B* 72 (2005) 045333.

97. G. Bastarg: *Wave Mechanics Applied to Semiconductor Heterostructures,* Les Éditions, Les Ulis (1990).

98. T. Mano. T. Kuroda, S. Sanquinetti, T. Ochiai, T. Tateno, J. S. Kim, T. Noda, M. Kawabe, K. Sakoda, G. Kido, N. Koguchi; *Nano Lett.* 5 (2005) 425.

99. H. Lee, R. Lowe-Webb, W. Yang, P.C. Sercel; *Appl. Phys. Lett.* 72 (1998) 812.

100. K. Watanabe, N. Koguchi, Y. Gotoh; *Jpn J Appl. Phys.* 39 (2000) L79.

101. T. Mano, T. Kuroda, M. Yamagiwa, G. Kido, K. Sakoda, N. Koguchi; *Appl. Phys. Lett.* 89 (2006) 183102.

102. G.A. Satunkin; J. Cryst. *Growth* 255 (2003) 170.

103. T. Mano, K. Watanabe, S. Tsukamoto, H. Fujikoa, M. Oshima, N. Koguchi; *J. Cryst. Growth* 209 (2000) 504.

104. M. Yamagiwa, T. Mano, T. Kuroda, T. Tateno, K. Sakoda, G. Kido, N. Koguchi; *Appl. Phys. Lett.* 89 (2006) 113115.

105. R. Pomraenke, C. Lienau, Y.I. Mazur, Z.H. Wang, B. Liang, G.G. Tarasov, G.J. Salamo; *Phys. Rev. B* 77 (2008) 075314.

106. R. Sonmuang, S. Kiravittaya, O.G. Schmidt; *Appl. Phys. Lett.* 82 (2003) 2892.

107. P.A. Gonzalez, J.M. Sanchez, Y. Gonalez, B. Alen, D. Fuster, L. Gonzalez; *Cryst. Growth Des.* 9 (2009) 2525.

108. K.A. Sablon, J.H. Lee, Z.M. Wang, J.H. Scultz, G.J. Salamo; *Appl. Ph ys. Lett.* 92 (2008) 203106.

109. M.K. Yakes, C.D. Cress, J.G. Tiscler, A.S. Bracker; *Nano* 4 (2010) 3877.

110. J.H. Lee, Zh.M. Wang, Yu, I. Mazur, G.J. Salamo; *Appl. Phys. Lett.* 89 (2006) 202101.

6

CHAPTER

THE PREPARATION AND APPLICATION OF DENDRIMER MODIFIED CDTE/CDS NEAR INFRARED QUANTUM DOTS FOR BRAIN CANCER CELLS IMAGING

Qingke Bai [1], Zhenguo Zhao [2,*], Haijing Sui [2], Juan Chen [1], Xiuhai Xie [2] and Feng Wen [2]

[1] *Department of Neurology, Pudong People's Hospital, Shanghai 201299, China*

[2] *Department of Radiology, Pudong People's Hospital, Shanghai 201299, China*

1. INTRODUCTION

As for brain imaging diagnosis, both magnetic resonance imaging (MRI) and positron emission tomography (PET) are the current gold standard imaging methods for brain cancer, but the functional mapping information on the human brain has not been not fully accumulated [1]. Furthermore, experimental environments for PET and MRI studies are quite different from the daily one, and most patients find this is stressful. Therefore, a new neuroimaging technique that is completely noninvasive and does not require strict motion restriction is desirable.

Near infrared (NIR) optical imaging, as a noninvasive optical technique, has emerged as an attractive imaging technology for brain imaging [2]. Previous studies revealed that the NIR molecule Cy5.5 were injected to rats, and the NIR imaging was perfectly companied with the size of true tumor after surgical exposure [3]. However, the NIR fluorescence of Cy5.5 is not stable in a biologically environment which is easy to bleach or dye *in vivo*. Compared with organic molecule, NIR QDs is more suitable for NIR optical *in vivo* imaging due to their unique optical advantages such as higher photoluminescence and photostability [3]. Moreover, the size of NIR QDs is normally less than 10 nm, which is enough small to cross the BBB [3,4].

To date, a number of NIR-emitting QDs, such as CdTe/CdS [5], CdHgTe [6], CdTeS [7] and CdTeSe [8] QDs, have been successfully synthesized. Recently, some studies reported the directly using dendrimer as a capping agent for preparing QDs could prevent QDs from further aggregation in water phrase [9,10]. Furthermore, dendrimer modified QDs enable to decrease the release of metal ions from this organic and inorganic hybrid architecture in biological environments [11].

In this study, the one-pot water-phrase synthesis of hybrid nanocomposites constituted by CdTe/CdS NIR QDs and different generations (from 3 up to 5) of plolyamidoamine dendrimers is presented. Moreover, the feasibility of using this nanocomposite as an imaging agent for labeling brain tumor cells is also assessed *in vitro*.

2. MATERIALS AND METHODS

2.1. Materials

Polypropylenimine dendrimer (generations 3 to 5), $CdCl_2$ (99%), Tellurium and selenium powder (99%) and NaBH4 were purchased from sigma Aldrich. Milli-Q water (Millipore, Mississauga, ON, Canada) was used as the solvent for all solutions.

2.2. Preparation of CdTe/CdS Core/Shell QDs

Dendrimer of different generations (G_n = 3–5) (10.0 mg) were dissolved in 20.0 mL in deionised water and stirred for 24 h. CdTe/CdS QDs were prepared in aqueous solution by a modified previous method. Typically, dendrimer was added to a N_2-saturated 2.5 mM $CdCl_2$ solution, followed by adjustment to the desired pH value (about 11) by the addition of 5 M NaOH solution. Then the precursor solution was subjected to microwave irradiation at 120 °C for about 2 min. Then, a portion of concentrated CdTe core QDs solution was mixed with a N_2-saturated CdS precursor solution. The mixed solution was subjected to microwave irradiation at 100 °C.

2.3. Structural and Optical Characterizations

TEM and HR-TEM images were recorded from a JEOL-3010 electron microscope (JEOL, Tokyo, Japan). Zeta potential and size distribution of our freshly prepared QDs in water solution measured by Nano ZS (Malvern, Worcestershire, UK). QDs powder (Christ Alpha, Shanghai, China) was used for XRD samples. UV-vis absorption spectra, PL spectra and Lifetime measurements were recorded using a Lambda 950 UV-vis spectrophotometer (Perkin Elmer, Wokingham, UK), F-4500 fluorescence spectrofluorimeter (Hitachi, Tokyo, Japan) and FL920-fluorescence lifetime spectrometer (Edinburgh Instruments, Edinburgh, UK), respectively.

For determination the fluorescence stability of dendrimer modified QDs, 1.0 mL dendrimer modified CdTe/CdS QDs solution and different amount of test solutions were sequentially added into test tubes, then diluted to the mark with water and mixed thoroughly. The resulting solutions were used to monitor the change of emission fluorescence intensity by means of spectrofluorimeter.

2.4. Cell Viability and Proliferation Assays

Human embryonic kidney cell line (HEK 293T), human brain cancer cell line (U251) and human gastric epithelial cell line (GES-1) were seeded at a density of 1×10^4 cells per well in the 96-well plates, and then added with 50 μL of PBS and various concentrations (0.5 nM, 5 nM and 50 nM) of GSH-CdTe/CdSe QDs. After the additional 24 h and 48 h incubation, cell survivals were measured using a tetrazolium salt MTT assay. Fresh growth medium (180 μL) and MTT (20 μL, 5 $mg \cdot mL^{-1}$) solution were added to each well. The plate was incubated for 4 h, and then 200 μL of dimethyl sulphoxide (DMSO) was added to each well to dissolve the purple formazan crystals. Finally, the absorbance at 492 nm of each well was measure.

2.5. Cell Imaging

Human brain cancer U251 cells were plated at a density of 5×10^5 cells per well in three 24-well plates with a glass cover slip bottom and incubated for 12 h at 37 °C with 5% CO_2. Dendrimer modified CdTe/CdS QDs were subsequently added to the wells for incubation 24 h at 37 °C. Then, these cells were fixed by 5% paraformaldehyde and stained with 4,6-diamidino-2-phenylindole (DAPI). The cell imaging was analyzed by using an inverted confocal microscope (Leica TCS SP2 AOBS, CNSI, Heidelberg, Germany).

3. RESULTS AND DISCUSSION

3.1. Characterization of Dendrimer Modified CdTe/CdS QDs

Figure 1 shows the TEM and HR-TEM images of the CdTe core (Figure 1a,b) and dendrimer modified CdTe/CdS core/shell (Figure 1d,e) QDs, which both the CdTe and dendrimer modified CdTe/CdS QDs displayed good dispersibility and water-solubility. As shown in

Figure 1c,f, the average size of the CdTe and dendrimer modified CdTe/CdS QDs is about 3.25 ± 0.2 nm and 4.05 ± 0.3 nm, respectively. Figure 1g shows the powder XRD pattern of the CdTe core and dendrimer modified CdTe/CdS core/shell QDs, and the XRD of CdTe/CdS QDs is matched previous reported results. Moreover, these main diffraction peaks of (111), (220) and (311) crystal plane also were indicated in SAED patterns (shown in Figure 1h). The hydro diameter and zeta potential of our products are shown in Figure 1i,j, which is about 20.3 nm and +9.73 eV, respectively.

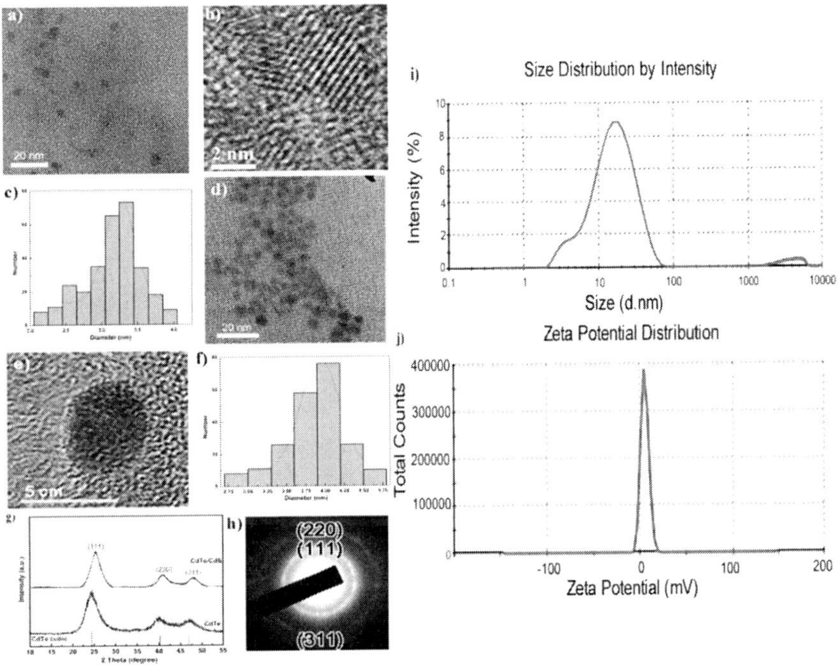

Figure 1. (**a**) Transmission electron microscopy (TEM) and (**b**) high-resolution (HR)-TEM images of CdTe QDs; (**c**) size distribution analysis of CdTe quantum dots (QDs); (**d**) TEM and (**e**) HR-TEM of dendrimer modified CdTe/CdS QDs; (**f**) size distribution analysis of dendrimer modified CdTe/CdS QDs; (**g**) X-ray powder diffraction (XRD) diffraction patterns of CdTe QDs and dendrimer modified CdTe/CdS QDs; (**h**) Selected area electron diffraction (SAED) pattern of dendrimer modified CdTe/CdS QDs; (**i**) zeta potential and (**j**) hydro diameter distribution of freshly prepared QDs.

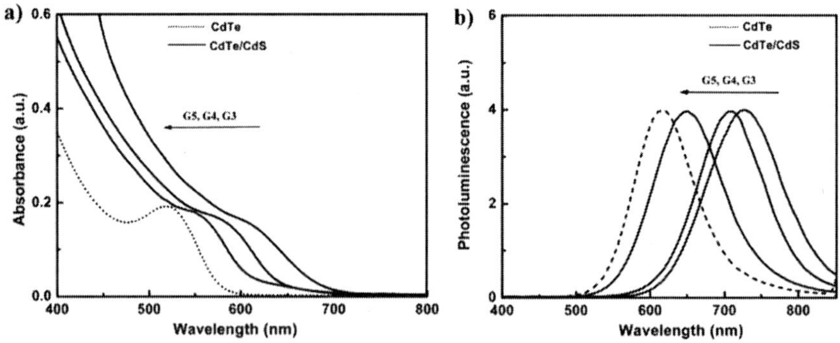

Figure 2. (**a**) UV-vis and Photoluminescence spectra of CdTe (dashed line) and CdTe/CdS QDs modified with different generations of dendrimer (solid lines); (**b**) Fluorescence decay curves of a CdTe core (dashed line) and CdTe/CdS QDs modified with different generations of dendrimer (solid lines).

As shown in Figure 2a, in the UV-vis absorption spectrum, there is a blue shift of the absorption band as the generation of the dendrimer is increased. The absorption shift that is observed for the different generations of dendrimers confirms that the amine groups are participating with dendrimers in the capping of the QDs. In the PL spectrum, the CdTe QDs exhibited emission at 562 nm with the excitation wavelength set of 365 nm. Compared to CdTe QDs, the emission wavelength of dendrimer modified core/shell CdTe/CdS QDs shifted to 650 nm which has been taken as a feature of QDs with spatial separated charge carriers [12]. Furthermore, the excitoinc peak of the core/shell CdTe/CdS QDs shifted to shorter wavelength companying with increasing the generations of dendrimers. This phenomenon of blue shift in emission peak is consistent with the known theory of quantum size effects.

3.2. The Stability of Dendrimer Modified CdTe/CdS QDs

As seen in Figure 3a, the photoluminescence (PL) lifetimes of freshly prepared CdTe QDs and dendrimer modified CdTe/CdS QDs is about

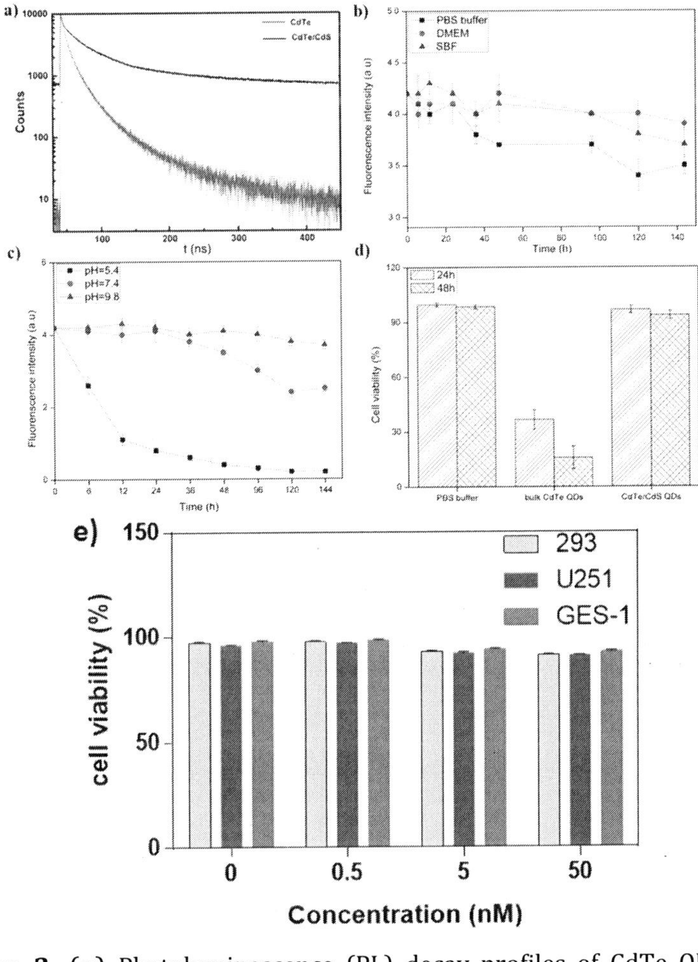

Figure 3. (**a**) Photoluminescence (PL) decay profiles of CdTe QDs and dendrimer modified CdTe/CdS QDs; (**b**) Effect of different buffer on the fluorescence intensity of dendrimer modified CdTe/CdS QDs at different time intervals; (**c**) Effect of different buffers on the fluorescence intensity of dendrimer modified CdTe/CdS QDs at different time intervals; (**d**) Cell viability of 293T cells after incubation with PBS buffer, bulk CdTe QDs and dendrimer modified CdTe/CdS QDs at 24 h and 48 h, respectively; (**e**) Cell viabilities of 293T cells, U251 cells and gastric epithelial cell line (GES-1) cells after incubating with various concentrations (0.5, 5, and 50 nM) of freshly prepared QDs for 24 h. The 0 nM concentration of QDs is taken as a control.

20.1 ns and 58.3 ns, respectively. The great increase of the decay lifetime of dendrimer modified CdTe/CdS QDs is ascribed to the spatial separation of electron and hole in type-II QDs [13]. Additionally, the PL quantum yields (QYs) of CdTe QDs and dendrimer modified CdTe/CdS QDs are 25% and 55%, respectively.

In order to develop a stable PL signal for the biological applications, the fluorescence stability of dendrimer modified CdTe/CdS QDs was accessed under various tested conditions such as pH, buffer solution and incubation time, respectively. As shown in Figure 3b, the PL signal of dendrimer-CdTe/CdS QDs is more stable when they were stored in Dulbecco's modified eagle medium (DMEM) and stimulate body fluid (SBF) buffer compared with they were stored in PBS buffer. Particularly, the fluorescence intensity of dendrimer modified CdTe/CdS QDs remains at 88% of the initial value after storage for 144 h in the PBS buffer. However, exposure of the QDs solutions to the acidic condition resulted in a decreased PL intensity, but it was basically stable under either neutral condition or alkaline condition (Figure 3c). This phenomenon is attributed to the surface charge characteristics of dendrimer modified QDs [14].

As shown in Figure 3d, the cytotoxicity of as-prepared dendrimer modified CdTe/CdS QDs was measured by MTT analysis. The cell viability of 293T cells treated with dendrimer modified CdTe/CdS QDs (50 nM) was about 95.2% and 93.2% after 24 h and 48 h incubation, while the cell viability was only about 36.7% after incubation 293T cells directly exposed with bulk CdTe QDs for 24 h, which proved that the dendrimer modified CdTe/CdS QDs had lower cytotoxicity than bulk QDs. Furthermore, none of cell lines including 293T, U251 and GES-1 cells shows a significant decrease in cell viability after exposing dendrimer modified CdTe/CdS QDs in the concentrations arrange from 0.5 to 50 mM for 24 h (Figure 3e). In a word, our prepared dendrimer modified CdTe/CdS QDs demonstrates an impressive biocompatibility with ultra-low toxicity, which is benefited to our following research.

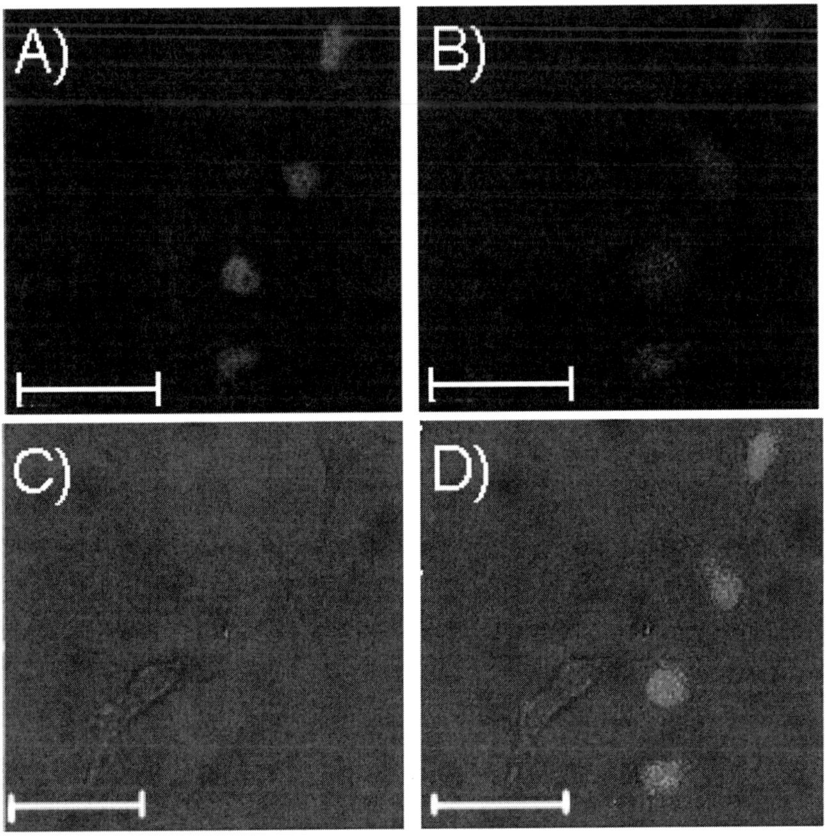

Figure 4. Fluorescence images of primary cultured U251 cells labelled with dendrime modified QDs. The U251 cells were stained by DAPI (Blue, **A**); Dendrimer modified CdTe/CdS QDs (Red, **B**); bright field (**C**) and merge images (**D**); Original magnification of (**A**–**D**) is 200×.

3.3. Cell Imaging

To ensure the feasibility of the NIRemitting dendrimer modified CdTe/CdS QDs for brain cancer imaging, we mainly applied them for brain tumor U251 cells imaging *in vitro*. As shown in Figure 4, we clearly observed the NIR emission from the cytoplasmic compartment of U251 cells. Moreover, the cell nucleus of U251 cells were hardly observed the NIR fluorescence of dendrimer modified CdTe/CdS QDs. This is probably due to the fact that the dendrimer modified QDs is easily interacted with cytosolic proteins during incubation, but is too large to enter the nucleus. This preliminary results show that dendrimer modified CdTe/CdS QDs with a suitable size and narrow, intense emission will be a promising probe for brain imaging *in vivo.*

4. CONCLUSIONS

In this work, we have had developed the synthesis procedure of dendrimer modified CdTe/CdS QDs in aqueous phase. Our data demonstrate that the as-prepared dendrimer modified CdTe/CdS QDs has a stable NIR fluorescence and favorable biocompatibility, which is vital for future biological applications. The as-prepared dendrimer modified CdTe/CdS QDs is also successfully applied for cell imaging *in vitro*. The dendrimer modified CdTe/CdS QDs will probably become an attractive alternative probe in future brain cancer imaging.

ACKNOWLEDGMENTS

This work is funded by the Pudong medical system point discipline foundation (PWZx2014-15), Pudong medical system leading talent foundation (PWR12010-02), Health Bureau of Shanghai (Grant No. 20134179), Research Grant for Health Science and Technology of Pudong Health Bureau of Shanghai (Grant No. PKJ2013-Y27), The 2013 Key Disciplines Group Construction Project of Pudong people's Hospital of Shanghai (Grant No. PRxkq2013-03).

AUTHOR CONTRIBUTIONS

Conceived and designed the experiments: Qingke Bai, Haijing Sui, Zhenguo Zhao, Feng Wen; Performed the experiments: Qingke Bai, Juan Chen, Xiuhai Xie; Analyzed the data: Qingke Bai, Haijing Sui, Zhenguo Zhao, Feng Wen; Contributed reagents/materials/analysis tools: Qingke Bai, Haijing Sui, Zhenguo Zhao, Feng Wen.

REFERENCES

1. Young, G.S.; Stauss, J.; Mukundan, S. Advanced Imaging of Adult Brain Tumors with MRI and PET. In *Neuro-Oncology: Blue Books of Neurology Series*; Saunders: Philadelphia, PA, USA, 2010; pp. 71–78.
2. Hoshi, Y. Functional Near-Infrared Optical Imaging: Utility and Limitations in Human Brain Mapping. *Psychophysiology* **2003**, *40*, 511–520.
3. Kircher, M.F.; Mahmood, U.; King, R.S.; Weissleder, R.; Josephson, L. A Multimodal Nanoparticle for Preoperative Magnetic Resonance Imaging and Intraoperative Optical Brain Tumor Delineation. *Cancer Res.* **2003**, *63*, 8122–8125.
4. Koo, Y.-E.L.; Reddy, G.R.; Bhojani, M.; Schneider, R.; Philbert, M.A.; Rehemtulla, A.; Ross, B.D.; Kopelman, R. Brain cancer diagnosis and therapy with nanoplatforms. *Adv. Drug Deliv. Rev.* **2006**, *58*, 1556–1577.
5. Zhao, D.; He, Z.; Chan, W.; Choi, M.M. Synthesis and characterization of high-quality water-soluble near-infrared-emitting CdTe/CdS quantum dots capped by *N*-acetyl-l-cysteine via hydrothermal method. *J. Phys. Chem. C* **2009**, *113*, 1293–1300.
6. Chen, H.; Wang, Y.; Xu, J.; Ji, J.; Zhang, J.; Hu, Y.; Gu, Y. Non-invasive near infrared fluorescence imaging of CdHgTe quantum dots in mouse model. *J. Fluores.* **2008**, *18*, 801–811.

7. Mao, W.; Guo, J.; Yang, W.; Wang, C.; He, J.; Chen, J. Synthesis of high-quality near-infrared-emitting CdTeS alloyed quantum dots via the hydrothermal method. *Nanotechnology* **2007**, *18*, 485–611.

8. Liang, G.-X.; Gu, M.-M.; Zhang, J.-R.; Zhu, J.-J. Preparation and bioapplication of high-quality, water-soluble, biocompatible, and near-infrared-emitting CdSeTe alloyed quantum dots. *Nanotechnology* **2009**, *20*.

9. Cong, R.-M.; Luo, Y.-J.; Li, G.-P.; Tan, H.-M. *In-situ* Synthesis of CdS-ZnS Core-shell Structure Quantum Dots Inside PAMAM Dendrimer Templates. *Chem. J. Chin. Univ.* **2006**, *27*, 793–796.

10. Yamamoto, D.; Koshiyama, T.; Watanabe, S.; Miyahara, M.T. Synthesis and photoluminescence characterization of dendrimer-encapsulated CdS quantum dots. *Colloids Surf. A Physicochem. Eng. Asp.* **2012**, *411*, 12–17.

11. Zhiming, L.; Peng, H.; Rong, H.; Xiaomin, Z.; Chenchen, B.; Qiushi, R.; Daxiang, C. Aptamer-conjugated dendrimer-modified quantum dots for glioblastoma cells imaging. *J. Phys. Conf. Ser.* **2009**, *188*, 012032.

12. Yu, K.; Zaman, B.; Romanova, S.; Wang, D.-S.; Ripmeester, J.A. Sequential synthesis of type II colloidal CdTe/CdSe core-shell nanocrystals. *Small* **2005**, *1*, 332–338.

13. Chang, J.-Y.; Wang, S.-R.; Yang, C.-H. Synthesis and characterization of CdTe/CdS and CdTe/CdSe core/shell type-II quantum dots in a noncoordinating solvent. *Nanotechnology* **2007**, *18*, 345602–345606.

14. Campos, B.B.; Algarra, M.; Alonso, B.; Casado, C.M.; Esteves da Silva, J.C.G. Mercury(ii) sensing based on the quenching of fluorescence of CdS-dendrimer nanocomposites. *Analyst* **2009**, *134*, 2447–2452

7

CHAPTER

SEMICONDUCTOR QUANTUM DOTS FOR BIOMEDICIAL APPLICATIONS

Lijia Shao, Yanfang Gao and Feng Yan *

Jiangsu Affiliated Cancer Hospital with Nanjing Medical University, Jiangsu Institute of Cancer Prevention and Cure, Nanjing 210009, China

ABSTRACT

Semiconductor quantum dots (QDs) are nanometre-scale crystals, which have unique photophysical properties, such as size-dependent optical properties, high fluorescence quantum yields, and excellent stability against photobleaching. These properties enable QDs as the promising optical labels for the biological applications, such as multiplexed analysis of immunocomplexes or DNA hybridization processes, cell sorting and tracing, in vivo imaging and diagnostics in biomedicine. Meanwhile, QDs can be used as labels for the electrochemical detection of DNA or proteins. This article reviews the synthesis and toxicity of QDs and their optical and electrochemical bioanalytical applications. Especially the application of QDs in biomedicine such as delivering, cell targeting and imaging for cancer research, and in vivo photodynamic therapy (PDT) of cancer are briefly discussed.

Keywords: quantum dots; bioanalysis; toxicology delivery; photodynamic therapy; cell imaging

1. INTRODUCTION

Quantum dots (QDs) as colloidal nanocrystalline semiconductors have unique photophysical properties due to quantum confinement effects. They emit different wavelengths over a broad range of the light spectrum from visible to infrared, depending on their sizes and chemical compositions. Compared with the traditional organic fluorophores (e.g., organic dyes and fluorescent proteins), QDs have unique optical and electronic properties, such as larger absorption coefficients, size-tunable light emission, superior signal brightness, resistance to photobleaching and simultaneous excitation of multiple fluorescence colors [1–6]. In addition, the large-surface area of QDs is beneficial to covalently link to biorecognition molecules, such as peptides, antibodies, nucleic acids or small-molecule ligands for further application as fluorescent probes (Figure 1).

2. THE SURFACE CHEMISTRY AND TOXICITY OF QDS

Early in the 1990s, Bawendi and coworkers first reported a synthesis protocol for QDs with highly monodisperse, regular core structure and tunable particle size [29,30]. Up to now, the most successful and well-developed method to prepare highly luminescent II–VI QDs is the TOP/TOPO synthetic approach [31]. However, these QDs are insoluble in water, which limits their biological applications. Therefore, a number of surface functionalization studies have been developed to make QDs water-soluble and biologically compatible [29–37].

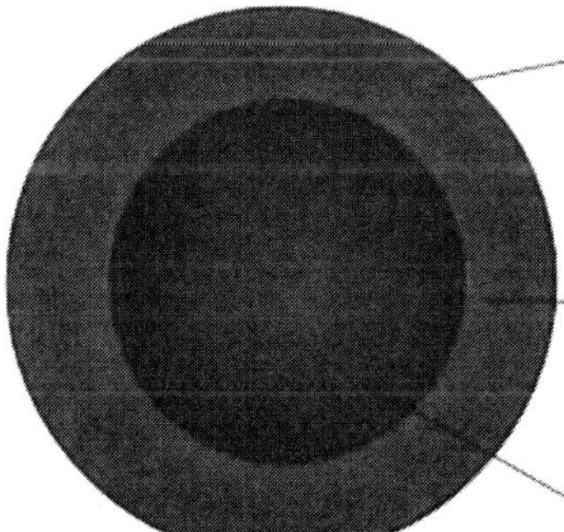

Figure 1. Schematic representation of a quantum dot. QDs are nanocrystals composed of a core of a semiconductor, usually composed of elements from groups II–IV, e.g., CdSe, or groups III–V, e.g., InP. The shell is typically a higher bandgap material such as ZnS. Finally, a capping outer layer such as silica can offer large-surface area for covalently linking to biorecognition molecules such as peptides, antibodies, nucleic acids and small-molecule ligands for further application. The diameter of QDs ranges between 2–10 nm.

These properties of QDs herald a revolution from electronic materials science to biological applications [7]. Current and projected applications of QDs include using fluorescent labels for cellular labeling [1,8,9], intracellular sensors [10,11], deep-tissue and tumor targeting and imaging agents [7,12–17], sensitizers for photodynamic therapy (PDT) [18–21], vectors for gene therapy [22–26], magnetic resonance imaging (MRI) contrast agents [27,28] and so on. This review mainly summarizes the development of synthesis, the surface modification and toxicity of QDs, and briefly focuses on the application developments of QDs in the biomedical field.

In one common approach, the original hydrophobic coatings are replaced by water-soluble functional molecules (e.g., dithiothreitol [38–40], mercaptocarbonic acids [41–44], 2-aminoethanethiol [33,45], dihydrolipoic acid [34–36,46,47], oligomeric phosphines [37,48], peptides [49–57], and cross-linked dendrons [58–61]) through the ligand exchange reactions. Because the optical properties of the inorganic core are often very sensitive to the surface, the ligand exchange process may result in poorer performance, particularly in the case of quantum dots [62].

The second approach is to encapsulate QDs in an amphiphile whose hydrophobic ends interleave with, but do not replace, the organic coating on QDs. This improvement for QDs synthesis is significant: (1) protecting the core/shell structure and maintaining the original photophysics of QDs; (2) making QDs water-soluble; (3) providing a biological interface and multiple functions [7]. However these kinds of QDs are not stable in biological settings because of relativelyweak anchoring of the single and double hydrophobic tails to the particle. Additionally, the hydrophilic end groups of even biocompatible surfactants may not protect nanocrystals from nonspecific biomolecular interactions [31]. Scientists have used amphiphilic polymers instead of simple amphiphile because single polymer chains can contain multiple hydrophobic units, their interactions with the native organic coatings on QDs can be numerous, and thus the encapsulant can be bound more strongly than conventional surfactants. However, the range of amphiphilic polymers for creating stable and nonaggregating QDs in biological settings has been relatively limited. Up to now, most of the amphiphilic polymers used are commercial and their hydrophobic/hydrophilic ratios are fixed, hence the cost is high and it may be different to control the process of forming water-soluble QDs and to optimize the forming conditions [31].

Although QDs have great prospects, the toxicity of QDs cannot be overlooked. During the processing of biological applications (e.g., cancer imaging, targeting and PDT treatment), the degradation products of QDs will do harm to the cells which they contact with, or produce immune responses with the components in blood [17]. The toxic degradation production routes are: first, the oxidation of the nanoparticle core/shell material can cause the release of free cadmium or other heavy metals, which will interrupt the normal cell activities [18]; secondly, the photosensitized production of reactive oxygen intermediates (ROI) also plays an important role in mediating the cell damage [63]; thirdly, the toxicity of capping materials should also be considered, several groups in capping materials such as mercaptoacetic acid and tri-n-octylphosphine oxide (TOPO) could produce toxicity to cells [12].

To reduce the cytotoxicity of QDs, replacement of the cadmium by nontoxic or less-toxic metals such as indium (In), or encapsulation of the core with a biocompatibile shell should be considered. Though In-based semiconducting dots contain arsenic, another toxin, the cytotoxicity of these dots may be small enough to keep the toxicity low. Fisher and coworkers [64] found that QDs could remain within the body for very long periods. Kim [8] reported that larger QDs generally accumulated in the reticuloendothelial system, such as the liver, spleen and lymphatic system for several months, but the size less than 5 nm could be removed by the kidney quickly. So in order to minimize the toxicity of QDs, QDs can be designed as smaller as they can, which can help them more easily to clean them out from the body.

In spite of the fact many investigators have paid close attention to and observed the side-effect of QDs, the definite metabolism of QDs in vivo remains uncertain [65–68]. Thus, it is still a necessary issue to investigate the detailed biochemical and pharmacological mechanism for further application of QDs in the human body.

3. DELIVERING QDS INTO CELLS

Effective delivery of QDs into the targeted-cell is the primary requirement for the bioapplications of QDs [9,15,17,20,32,54]. It is a major step because if QDs cannot reach their site of action in vivo, they is useless. Furthermore, efficient delivery can also allow a reduction in dosage level, avoid non-specific side effects and reduce toxicity risks [66,69,70]. The current methods for delivering QDs into cells mainly include passive delivery, facilitated delivery and active delivery.

The general passive delivery for QDs is endocytosis, which is simple, without further functionalization of the QDs surface with a targeting ligand for uptake [66]. By incubating with the cells at appropriate concentration and exposure time, QDs will enter into cells though the nonspecific cell endocytosis. However, the nonspecific ingestion of this mode caused ineffective endosomal escape, and would impede the delivery of QDs to the cytoplasm or other organelles. Furthermore, high intracellular concentration of QDs can enhance the cytotoxicity in some cases [69].

Facilitated delivery includes four ways: peptide-mediated uptake, protein-mediated delivery, polymer-mediated delivery and small molecule-mediated delivery [66]. Generally, these molecules are noncovalently assembled onto the surface of QDs for bioconjugation. Facilitated delivery could reduce the nonspecific absorption and side effects. However, QDs could also be uptaken by cell through endocytosis, leading to endosomal sequestration during the facilitated delivery strategies (Figure 2). As is well known, the high acidic of endosomes could degrade the QDs conjugates over time, thus free Polyethyleneimine (PEI) was used to encapsule the QDs conjugates to increase the stability [70]. Considering further application of cell imaging, more general endosomal escape strategies need to be developed in order to expand the application of facilitated delivery.

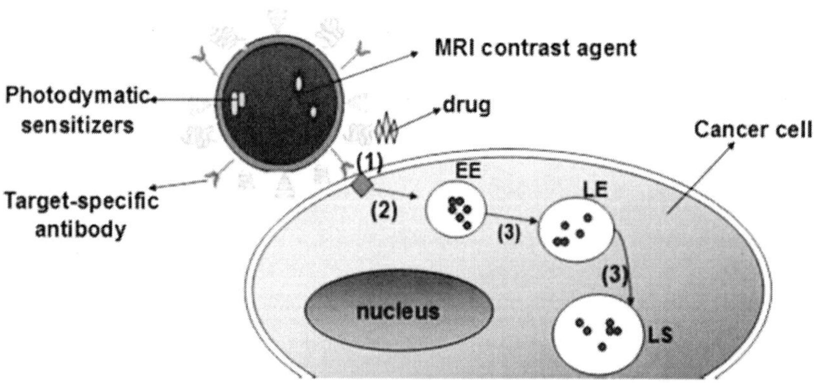

Figure 2. Schematic representation of delivering QDs into cells, the process comprises of three major stages: (1) endocytosis; (2) sequestering in early endosome (EE); (3) translocation to later endosomes (LE) or lysosomes (LS).

Table 1. Selected strategies for the intracellular delivery of QDs.

Strategy	Mechanism	Examples	Targeted Cells	References
Passive uptake	Electrostatic interactions	-	HeLa	[15]
			Human macrophages	[27,72]
			Breast cancer (MDA-MB-231)	[73,74]
			Human melanoma cells (LU1205)	[75]
Facilitated delivery	Peptide-mediated	TAT	Human embryonic kidney	[76]
			HeLa	[77]
			Mesenchymal stem cells	[53]
			Jurkat cells	[28]
		Pep-1 (Chariot)	Osteoblast	[78]
			Vascular endothelial cells	[78]
		RGD motify	Fibroblast (NIH 3T3)	[79]
			Epidermoid carcinoma	[80]
	Protein-mediated	Neuropeptide	HeLa	[17]
		Transferrin	Human pancreatic cancer	[2,81,82]
		Antibody	Breast cancer (MCF-7)	[11]
		EGF	Mesenchymal stem cells	[83]
			Chinese hamster ovary	[84-86]
			Medulloblastoma tumors	[87]
			Glioma tumors	[87]
		Cholera toxin B	Fibroblast	[88,89]
		NGF	PC12 neural cells	[90,91]
	Polymer/lipid-mediated	Lipid polymers	Mouse lymphoma	[92]
			HeLa	[93]
			A549 epithelial lung HeLa	[94]
		Polyethylenei-mine	HeLa	[95]
	Drug-mediated	Tiopronin	Fibroblast	[96]
	Small molecule	Glucose/sugar	*S. cerevisiae* (Baker's yeast)	[97,98]
		Folate	Epidermal carcinoma	[99]
		Adenine/AMP	Bacteria (*Bacillus subtilis, E. coli*)	[100,101]
		Dopamin	A9 mouse fibroblast with transfected dopamine receptor	[102]
Active Delivery	Electroporation	-	HeLa	[71]
			Mouse neural stem progenitor cells	[9]
	Microinjection	-		[71]
			Xenopus embryo	[103]
			HeLa	[104]
			Human embryonic kidney	

Active delivery is a direct physical manipulation of the cell by electroporation and microinjection. In comparison to facilitated delivery, QDs conjugates are delivered directly to the cytoplasm via electroporation by an endocytic pathway, without subsequent endosomal escape. However, the high cellular mortality rate and intracellar aggregation occurring during the delivery should be conquered [71]. Compared with electroporation, microinjection could deliver the QDs directly to the cytoplasm with lower cell death rate, and the rate of microinjection of QDs conjugates to cells depends on the physical constraints of cells, including morphology, membrane thickness, height, etc. [66]. Furthermore, this technology is very expensive. Therefore, considering the coexistence of advantages and drawbacks of the mentioned approaches, the appropriate way for delivering QDs into cell should be determined according to the specific experimental requirements. The relationship between the specific examples and the delivery strategies are listed in Table 1 [66].

4. QDS-BASED CANCER TARGETING AND IMAGING

The photoluminescence (PL) of QDs is exceptionally bright and stable, making them potential candidates for biomedical imaging and therapeutic interventions. QDs conjugated with cancer specific ligands/antibodies/peptides were found to be effective for detecting and imaging human cancer cells. Gao and coworkers [67] firstly reported the QDs-antibody conjugates for in vivo targeting and imaging cancer, in which QDs-antibody conjugates were used as imaging probe for investigating and tracing QDs-PSMA antibody conjugates in mouse bearing subcutaneous human prostate cancer. It was found that the QDs-antibody conjugates were efficiently and uniformly distributed in prostate tumors due to the specific binding between PSMA antigen in prostate cancer cells and PSMA antibody on QDs. Cai and coworkers [105] conjugated NIR QDs with RED peptide, which could bind to the over-expressed αvβ3 integrin on the surface of U87MG glioblastoma cells and MDA-MB-435 human breast cancer cells to target cancer cell in vivo. By linking QDs to AFP (alpha-fetoprotein) antibody, an important marker for hepatocellular carcinoma cell lines, a specific immunofluorescent probes was obtained for further detection of AFP antibody in human serum. Yu et al. [106] demonstrated that the probe could target the specific hepatocellular carcinoma cells, and the expected results was obtained by investigating distribution of the probes in cancer cells by using a site-by-site measurement. Weng et al. [107] functionalized QDs with anti-HER2 scFv to synthesize the immunoliposome-based nanoparticles (QD-ILs). After incubating with HER2-overexpressing

SK-BR-3 and MCF-7/HER2 cells, the QD-ILs exhibited efficient receptor-mediated endocytosis. In vivo fluorescence imaging showed that QD-ILs had localized prominently in tumors as well as in MPS organs (Figure 3). Liu et al. [68] reported a QDs-based wavelength-resolved spectral imaging for molecular mapping of tumor heterogeneity on human prostate cancer tissue specimens. By conjugating different QDs with specific protein biomarkers, such as E-cadherin, high-molecular-weight cytokeratin, p63, and α-methylacyl CoA racemase, structural distinct prostate glands and single cancer cells could be detected and characterized within the complex microenvironments of radical prostatectomy and needle biopsy tissue specimens using the wavelength-resolved spectral imaging.

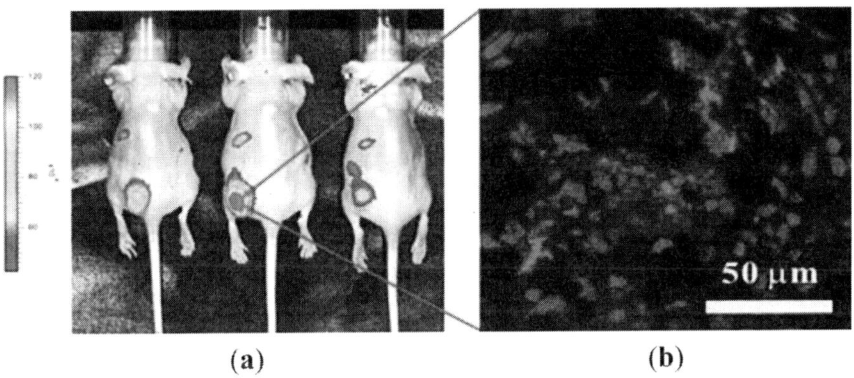

Figure 3. (a) Left panel: In vivo fluorescence imaging of three nude mice bearing MCF-7/HER2 xenografts implanted in the lower back 30 h after i.v. injection with anti-HER2 QD-ILs; (b) Right panel: A 5 μm section cut from frozen tumor tissues harvested at 48 h postinjection and examined by confocal microscopy by a 63× oil immersion objective (image size, 146 μm × 146 μm). The tumor section was examined in two-color scanning mode for nuclei stained by DAPI (blue) and QD-ILs (red). (Cited from Weng et al. [107]).

The main advantage of QDs imaging is that it is non-ionizing and less hazardous [108]. In recent years, several groups have used QD probes for fluorescence immunostaining of fixed cells and tissue specimens [109–113]. QD-based immunohistochemistry (IHC) can improve both diagnostic sensitivity and specificity. In addition, because multiplexed QD staining can be carried out on intact cells and tissue specimens, it is expected to provide correlated molecular and morphological information, at the same time, this type of integrated biomarker and morphological data are not available from traditional analytical methods such as mass spectrometry, gene chips, protein microarrays, and polymerase chain reactions [109]. However, medical applications of QD-based IHC have achieved only limited success. A major bottleneck is the lack of robust protocols to define the key parameters and steps [109]. For example, there are no consensuses on methods for QD-antibody (QD-Ab) bioconjugation, tissue specimen preparation, multicolor QD staining, image processing and data quantification. So it is necessary to solve these problems, and let the QDs move further.

5. QDS RELATED PHOTODYNAMIC THERAPY FOR CANCER

Presently, the conventional types of cancer treatment (chemotherapy and radiation therapy), work by destroying fast-growing cells, but other types of fast-growing healthy cells (such as blood and hair cells) also can be damaged along with cancer cells, causing adverse reactions, or side effects. These side effects can range from fatigue and flu-like symptoms to hair loss and blood clotting problems. PDT developed in last century has become an FDA-approved therapy for different malignancies and with potential in other ailments such as coronary heart disease, AIDS and psoriasis [63].

Exploration of the use of light-activated drugs known as photosensitizers (PS) has been one of the most active areas of photomedical research in recent years [18–21,63,114,115]. PDT uses the combination of a photosensitizing drug and light in the presence of oxygen to cause selective damage to the targeting tissue. During PDT, reactive oxygen intermediates (ROI) is generated in the diseased cells by a simple and controllable light-activated process, which involves a photosensitizer that is capable of absorbing light appropriate wavelength and transfers energy or electron to oxygen or other molecules, and creates ROI such as singlet oxygen (1O2), hydroxyl radical (OH), super oxide anion (O2−) and hydrogen peroxide (H2O2). Then ROI will immediately react with vital biomolecules in cell organelles, leading to cell damage, mutation, death and photooxidation of cell constituents [19,20,63,114,115]. Singlet oxygen (1O2) is regarded as the main mediator of photo-induced cytotoxicity in PDT, which causes oxidation and degradation of cellular components, and ultimately cell apoptosis. [20,63,114,115] (Figure 4).

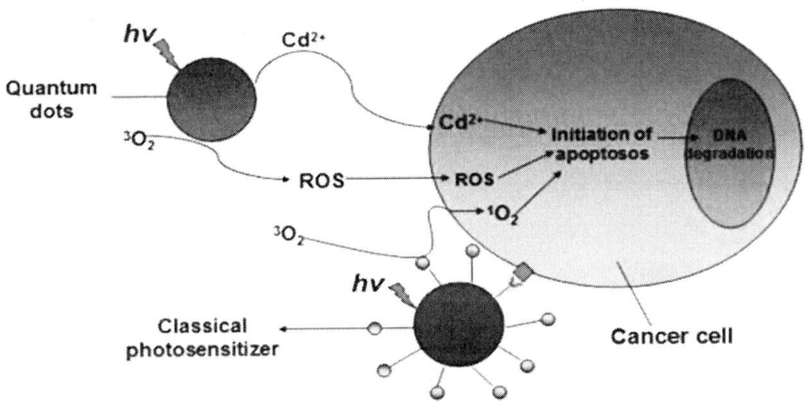

Figure 4. Schematic representation of possible mechanisms for induction of PDT processes by QDs and the classical photosensitizer.

The standard PS drugs for PDT are porphyrin, phthalocyanines and chlorine derivatives. Porphyrin derivatives are the first generation photosensitizer. Despite the clinical success of porphyrin derivatives, some of their disadvantages like prolonged cutaneous photosensitivity, chemical impurity and weak absorption at therapeutic wavelengths have inspired the development of new PDT photosensitizers with improved optical and chemical properties. Phthalocyanines derivatives have favorable photophysical and chemical properties, which include strong absorbance at long wavelengths and chemical tunability through substituent addition on the periphery of the macrocycle or on the axial ligands. However, like most photosensitizing agents, these PS have poor solubility in water and tend to aggregate in aqueous solutions, which can result in loss of photochemical activity and affect their cell penetrating properties [63]. To resolve such issues nanoparticles are currently being explored as potential delivery systems for PDT photosensitizers or directly as PDT agents. The novel QDs-PS conjugates are used as a high ratio of PDT agents and anticancer targeting antibodies, where QDs can act as nanoscaffolds and solubilizers. They can also function as "energy-harvesting antenna" for PDT therapy due to their large one- or two-photon absorption cross-sections. Thus, QDs can be efficiently exited even deep within tissues and sensitized proximal PDT agents via energy transfer from QDs to PDT [21].

The novel QDs-PS conjugates showed many advantages over conventional PS drugs [17–21,63,114,115]: (1) they are species with well-defined size, shape, and composition, and can be synthesized by relatively simple and inexpensive methods; (2) they have been shown to be nontoxic in the absence of light but have the potential to be cytotoxic under irradiation; (3) they have photostability, and tunable and strong absorption, which can be tuned from the UV their composition and size; (4) the surface coating of QDs can be modified to enable them to become water soluble, biocompatible and target-specific.

However, researchers should be further investigated on the basis of predominances of the QDs-PS compared to the convention PS drugs. Despite many desirable properties of QDs for PDT, there still remain several important issues that need to be addressed to fully assess their applicability as PS in PDT. One major issue is the toxicity profile of the QDs inside the cells and their overall photostability once exposed to biological environments [63]. Another important matter that should be carefully investigated is how their surface composition affects the photosensitization process. Still, QDs-PS conjugates for cancer therapy are only suitable to superficial tumours is also need to be resolved [18].

6. CONCLUSIONS AND OUTLOOK

In the last decade, the unique photophysical properties and functions of QDs have been widely investigated, making them one of the most promising nanomaterials. Their outstanding performances such as high fluorescence yields, stability against photobleaching and the size-dependent luminescence features of QDs provide broad variety of applications for QDs in many fields. By acting as fluorescent, and photoelectrochemical as well as electrochemical probes, various QDs-based optical and electrochemical bioanalysis have already been successfully explored for sensing a wide range of molecules with high sensitivity and specificity. Furthermore, as a biomedical label, QDs can make a worthy contribution to the development of new diagnostic and delivery systems due to their unique optical properties. By combination of functional biomolecule-nanoparticle hybrid systems and the optical imaging and biophysics, QDs have been used as optical reporter units of biocatalytic transformations and can probe intracellular processes in vitro. QDs as a novel probe for in vivo analysis and clinic therapy such as PDT open an attractive new field with promising prospectives in biomedicine.

ACKNOWLEDGMENTS

This review was financially supported by the National Basic Research Program of China (2010CB732400), the National Natural Science Foundation of China (21075055), the Program for Six Peak Talents of Jiangsu province (66) and the Leading Medical Talents Program from Department of Health of Jiangsu Province.

REFERENCES

1. Bruchez, M.J.; Moronne, M.; Gin, P.; Weiss, S.; Alivisatos, A.P. Semiconductor nanocrystals as fluorescent biological labels. Science 1998, 281, 2013–2016.

2. Chan, W.C.; Nie, S. Quantum dot bioconjugates for ultrasensitive nonisotopic detection. Science 1998, 281, 2016–2018.

3. Chan, W.C.; Maxwell, D.J.; Gao, X.H.; Bailey, R.E.; Han, M.Y.; Nie, S. Luminescent quantum dots for multiplexed biological detection and imaging. Curr. Opin. Biotechnol 2002, 13, 40–46.

4. Dabbousi, B.O.; Rodriguez-Viejo, J.; Mikulec, F.V.; Heine, J.R.; Mattoussi, H.; Ober, R.; Jensen, K.F.; Bawendi, M.G. (CdSe) ZnS core-shell quantum dots: Synthesis and characterization of a size series of highly luminescent nanocrystallites. J. Phys. Chem 1997, 101, 9463–9475.

5. Lim, Y.T.; Kim, S.; Nakayama, A.; Stott, N.E.; Bawendi, M.G.; Frangioni, J.V. Selection of quantum dot wavelengths for biomedical assays and imaging. Mol. Imaging 2003, 2, 50–64.

6. Mattoussi, H.; Kuno, M.K.; Goldman, E.R.; George, P.; Mauro, J.M. Colloidal semiconductor quantum dot conjugates in biosensing. In Optical Biosensors: Present and Future; Ligler, F.S., Rowe Taitt, C.A., Eds.; Elsevier: Amsterdam, The Netherlands, 2002; Chapter 17; pp. 537–569.

7. Michalet, X.; Pinaud, F.F.; Bentolila, L.A. Quantum dots for live cells, in vivo imaging, and diagnostics. Science 2005, 307, 538–544.

8. Kim, S.; Lim, Y.T.; Soltesz, E.G.; Grand, A.M.; Lee, J.; Nakayama, A. Near-infrared fluorescent type II quantum dots for sentinel lymph node mapping. Nat. Biotechnol 2004, 22, 93–97.

9. Chen, F.Q.; Gerion, D. Fluorescent CdSe/ZnS nanocrystal-peptide conjugates for long-term, nontoxic imaging and nuclear targeting in living cells. Nano Lett 2004, 4, 1827–1832.

10. Somers, R.C.; Bawendi, M.G.; Nocera, D.G. Nanocrystals as sensors. Green Chem 2007, 9, 403–410.

11. Qian, J.; Yong, K.T.; Roy, I.; Ohulchanskyy, T.Y.; Bergey, E.J.; Lee, H.H.; Tramposch, K.M.; He, S.; Maitra, A.; Prasad, P.N.

Imaging pancreatic cancer using surface-functionalized quantum dots. J. Phys. Chem 2007, 111, 6969–6972.

12. Smith, A.M.; Duan, H.; Mohs, A.M.; Nie, S. Bioconjugated quantum dots for in vivo molecular and cellular imaging. Adv. Drug Deliv. Rev 2008, 60, 1226–1240.

13. Medintz, I.L.; Uyeda, H.T.; Goldman, E.R.; Mattoussi, H. Quantum dot bioconjugates for imaging, labelling and sensing. Nat. Mater 2005, 4, 435–446.

14. Jaiswal, J.K.; Mattoussi, H.; Mauro, J.M.; Simon, S.M. Long-term multiple color imaging of live cells using quantum dot bioconjugates. Nat. Biotechnol 2003, 21, 47–51.

15. Bakalova, R.; Zhelev, Z.; Aoki, I.; Kanno, I. Designing quantum-dots probes. Nat. Photonics 2007, 1, 487–489.

16. Baron, R.; Willner, B.; Willner, I. Biomolecule-nanoparticle hybrids as functional units for nanobiotechnology. Chem. Commun 2007, 18, 323–332.

17. Biju, V.; Muraleedharan, D.; Nakayama, K.; Shinohara, Y.; Itoh, T.; Baba, Y.; Ishikawa, M. Quantum dot-insect neuropeptide conjugates for fluorescence imaging, transfection, and nucleus targeting of living cells. Langmuir 2007, 23, 10254–10261.

18. Juzenas, P.; Chen, W.; Sun, Y.P.; Coelho, M.; Generalov, R.; Generalova, N.; Christensen, I.L. Quantum dots and nanoparticles for photodynamic and radiation therapy of cancer. Adv. Drug Deliv. Rev 2008, 60, 1600–1614.

19. Ochsner, M. Photophysical and photobiological processes in the photodynamic therapy of tumours. J. Photochem. Photobiol. B 1997, 39, 1–18. Oleinick, N.L.; Evans, H.H. The photobiology of photodynamic therapy: Cellular targets and mechanisms. Radiat. Res 1998, 150, 146–156.

20. Tardivo, J.P.; Giglio, A.D.; Oliveira, C.S.; Gabrielli, D.S.; Junqueira, H.C.; Tada, D.B.; Severino, D.; Turchiello, R.F.; Baptista, M.S. Methylene blue in photodynamic therapy: From basic mechanisms to clinical applications. Photodiagn. Photodyn. Ther 2005, 2, 175–191.

21. Jiang, Z.; Li, R.; Todd, N.W.; Stass, S.A.; Jiang, F. Detecting genomic aberrations by fluorescence in situ hybridization with quantum dots-labeled probes. J. Nanosci. Nanotechnol 2007, 7, 4254–4259.

22. Burke, R.S.; Pun, S.H. Extracellular barriers to in vivo PEI and PEGylated PEI polyplex-mediated gene delivery to the liver. Bioconjug. Chem 2008, 19, 693–704.

23. Delehanty, J.B.; Boeneman, K.; Bradburne, C.E.; Robertson, K.; Medintz, I.L. Quantum dots: A powerful tool for understanding the intricacies of nanoparticle-mediated drug delivery. Expert Opin. Drug Deliv 2009, 6, 1091–1112.

24. Zhao, Y.; Zhao, L.; Zhou, L.; Zhi, Y.; Xu, J.; Wei, Z.; Zhang, K.X.; Ouellette, B.F.; Shen, H. Quantum dot conjugates for targeted silencing of bcr/abl gene by RNA interference in human myelogenous leukemia K562 cells. J. Nanosci. Nanotechnol 2010, 10, 5137–5143.

25. Bolhassani, A.; Safaiyan, S.; Rafati, S. Improvement of different vaccine delivery systems for cancer therapy. Mol. Cancer 2011.

26. Nabiev, I.; Mitchell, S.; Davies, A.; Williams, Y.; Kelleher, D.; Moore, R.; Gun'ko, Y.K.; Byrne, S.; Rakovich, Y.P.; Donegan, J.F.; Sukhanova, A.; Conroy, J.; Cottell, D.; Gaponik, N.; Rogach, A.; Volkov, Y. General strategy for designing functionalized magnetic microspheres for different bioapplications. Nano Lett 2007, 7, 3452–3461.

27. Lewin, M.; Carlesso, N.; Tung, C.H.; Tang, X.W.; Cory, D.; Scadden, D.T.; Weissleder, R. Tat peptide-derivatized magnetic nanoparticles allow in vivo tracking and recovery of progenitor cells. Nat. Biotechnol 2000, 18, 410–414.

28. Dabbousi, B.O.; Murray, C.B.; Rubner, M.F.; Bawendi, M.G. Langmuir-Blodgett manipulation of size-selected CdSe nanocrystallites. Chem. Mater 1994, 6, 216–219.

29. Danek, M.; Jensen, K.F.; Murray, C.B.; Bawendi, M.G. Electrospray organometallic chemical vapor deposition-A

novel technique for preparation of II–VI quantum dot composites. Appl. Phys. Lett 1994, 65, 2795–2797.

30. Wu, Z.; Zhao, Y.; Qiu, F.; Li, Y.; Wang, S.; Yang, B.; Chen, L.; Sun, J.; Wang, J. Forming water-soluble CdSe/ZnS QDs using amphiphilic polymers, stearyl methacrylate/methylacrylate copolymers with different hydrophobic moiety ratios and their optical properties and stability. Colloids Surf. A Physicochem. Eng. Asp 2009, 350, 121–129.

31. Ghasemi, Y.; Peymani, P.; Afifi, S. Quantum dot: Magic nanoparticle for imaging, detection and targeting. Acta Biomed 2009, 80, 156–165.

32. Lee, C.M.; Jang, D.; Cheong, S.J.; Kim, E.M.; Jeong, M.H.; Kim, S.H.; Kim, D.W.; Lim, S.T.; Sohn, M.H.; Jeong, H.J. Surface engineering of quantum dots for in vivo imaging. Nanotechnology 2010.

33. Uyeda, H.T.; Medintz, I.L.; Jaiswal, J.K.; Simon, S.M.; Mattoussi, H. Synthesis of compact multidentate ligands to prepare stable hydrophilic quantum dot fluorophores. J. Am. Chem. Soc 2005, 127, 3870–3878.

34. Pons, T.; Uyeda, H.T.; Medintz, I.L.; Mattoussi, H. Hydrodynamic dimensions, electrophoretic mobility, and stability of hydrophilic quantum dots. J. Phys. Chem. B 2006, 110, 20308–20316.

35. Smith, A.M.; Nie, S. Minimizing the hydrodynamic size of quantum dots with multifunctional multidentate polymer ligands. J. Am. Chem. Soc 2008, 130, 11278–11279.

36. Kim, S.; Bawendi, M.G. Oligomeric ligands for luminescent and stable nanocrystal quantum dots. J. Am. Chem. Soc 2003, 125, 14652–14653.

37. Stewart, M.H.; Susumu, K.; Mei, B.C.; Medintz, I.L.; Delehanty, J.B.; Blanco-Canosa, J.B.; Dawson, P.E.; Mattoussi, H. Multidentate poly(ethylene glycol) ligands provide colloidal stability to semiconductor and metallic nanocrystals in extreme conditions. J. Am. Chem. Soc 2010, 132, 9804–9813.

38. Kong, X.L.; Qi, H.; Zhou, H.X.; Ren, L.L.; Deng, C.Y.; Li, F.R. A novel sensitive immunoassay by nucleic acid barcode dot and its application in the detection of prostate-specific antigen. Clin. Chem. Lab. Med 2010, 48, 279–283.

39. Thangadurai, P.; Balaji, S.; Manoharan, P.T. Surface modification of CdS quantum dots using thiols-structural and photophysical studies. Nanotechnology 2008, 19, 435708:1–435708:8. [Google Scholar]

40. Yuan, Z.; Zhang, A.; Cao, Y.; Yang, J.; Zhu, Y.; Yang, P. Effect of mercaptocarboxylic acids on luminescent properties of CdTe quantum dots. J. Fluoresc 2011.

41. Petkar, K.C.; Chavhan, S.S.; Agatonovik-Kustrin, S.; Sawant, K.K. Nanostructured materials in drug and gene delivery: A review of the state of the art. Crit. Rev. Ther. Drug Carr. Syst 2011, 28, 101–164. [Google Scholar]

42. Zhao, L.; Pang, X.; Adhikary, R.; Petrich, J.W.; Jeffries-El, M.; Lin, Z. Organic-inorganic nanocomposites by placing conjugated polymers in intimate contact with quantum rods. Adv. Mater 2011, 23, 2844–2849.

43. Song, S.; Qin, Y.; He, Y.; Huang, Q.; Fan, C.; Chen, H.Y. Functional nanoprobes for ultrasensitive detection of biomolecules. Chem. Soc. Rev 2010, 39, 4234–4243.

44. Zorn, M.; Bae, W.K.; Kwak, J.; Lee, H.; Lee, C.; Zentel, R.; Char, K. Quantum dot-block copolymer hybrids with improved properties and their application to quantum dot light-emitting devices. ACS Nano 2009, 3, 1063–1068. [Google Scholar]

45. Goldman, E.R.; Mattoussi, H.; Anderson, G.P.; Medintz, I.L.; Mauro, J.M. Fluoroimmunoassays using antibody-conjugated quantum dots. Methods Mol. Biol 2005, 303, 19–34.

46. Liu, W.; Howarth, M.; Greytak, A.B.; Zheng, Y.; Nocera, D.G.; Ting, A.Y.; Bawendi, M.G. Compact biocompatible quantum dots functionalized for cellular imaging. J. Am. Chem. Soc 2008, 130, 1274–1284.

47. Sheng, W.; Kim, S.; Lee, J.; Kim, S.W.; Jensen, K.; Bawendi, M.G. In situ encapsulation of quantum dots into polymer microspheres. Langmuir 2006, 22, 3782–3790.

48. Bolhassani, A. Potential efficacy of cell-penetrating peptides for nucleic acid and drug delivery in cancer. Biochim. Biophys. Acta 2011, 1816, 232–246.

49. Wang, J.; Xia, J. Preferential binding of a novel polyhistidine peptide dendrimer ligand on quantum dots probed by capillary electrophoresis. Anal. Chem 2011, 83, 6323–6329. [Google Scholar]

50. Li, Y.; Zhou, Y.; Wang, H.Y.; Perrett, S.; Zhao, Y.; Tang, Z.; Nie, G. Chirality of glutathione surface coating affects the cytotoxicity of quantum dots. Angew. Chem. Int. Ed. Engl 2011, 50, 5860–5864.

51. Iyer, G.; Pinaud, F.; Xu, J.; Ebenstein, Y.; Li, J.; Chang, J.; Dahan, M.; Weiss, S. Aromatic aldehyde and hydrazine activated Peptide coated quantum dots for easy bioconjugation and live cell imaging. Bioconjug. Chem 2011, 22, 1006–1011.

52. Ranjbarvaziri, S.; Kiani, S.; Akhlaghi, A.; Vosough, A.; Baharvand, H.; Aghdami, N. Quantum dot labeling using positive charged peptides in human hematopoetic and mesenchymal stem cells. Biomaterials 2011, 32, 5195–5205.

53. Kuo, C.W.; Chueh, D.Y.; Singh, N.; Chien, F.C.; Chen, P. Targeted nuclear delivery using peptide-coated quantum dots. Bioconjug. Chem 2011, 22, 1073–1080.

54. Sapsford, K.E.; Granek, J.; Deschamps, J.R.; Boeneman, K.; Blanco-Canosa, J.B.; Dawson, P.E.; Susumu, K.; Stewart, M.H.; Medintz, I.L. Monitoring botulinum neurotoxin a activity with peptide-functionalized quantum dot resonance energy transfer sensors. ACS Nano 2011, 5, 2687–2699.

55. Liu, H.Y.; Gao, X. Engineering monovalent quantum dot-antibody bioconjugates with a hybrid gel system. Bioconjug. Chem 2011, 22, 510–517.

56. Jin, T.; Tiwari, D.K.; Tanaka, S.; Inouye, Y.; Yoshizawa, K.; Watanabe, T.M. Antibody-protein A conjugated quantum dots

for multiplexed imaging of surface receptors in living cells. Mol. Biosyst 2010, 6, 2325–2331.

57. Das, A.; Sanjayan, G.J.; Kecskés, M.; Yoo, L.; Gao, Z.G.; Jacobson, K.A. Nucleoside conjugates of quantum dots for characterization of G protein-coupled receptors: Strategies for immobilizing A2A adenosine receptor agonists. J. Nanobiotechnol 2010.

58. Zhao, Y.; Li, Y.; Song, Y.; Jiang, W.; Wu, Z.; Wang, Y.A.; Sun, J.; Wang, J. Architecture of stable and water-soluble CdSe/ZnS core-shell dendron nanocrystals via ligand exchange. J. Colloid. Interface Sci 2009, 339, 336–343.

59. Hwang, S.H.; Moorefield, C.N.; Wang, P.; Jeong, K.U.; Cheng, S.Z.; Kotta, K.K.; Newkome, G.R. Dendron-tethered and templated CdS quantum dots on single-walled carbon nanotubes. J. Am. Chem. Soc 2006, 128, 7505–7509.

60. Advincula, R.C. Hybrid organic-inorganic nanomaterials based on polythiophene dendronized nanoparticles. Dalton Trans 2006, 23, 2778–2784.

61. Rogach, A.L.; Nagesha, D.; Ostrander, J.W.; Giersig, M.; Kotov, N.A. "Raisin bun"-type composite spheres of silica and semiconductor nanocrystals. Chem. Mater 2000, 12, 2676–2685.

62. Samia, A.C.; Dayal, S.; Burda, C. Quantum dot-based energy transfer: Perspectives and potential for applications in photodynamic therapy. Photochem. Photobiol 2006, 82, 617–625.

63. Fischer, H.; Liu, L.; Pang, K.S.; Chan, W. Pharmacokinetics of nanoscale quantum dots: In vivo distribution, sequestration, and clearance in the rat. Adv. Funct. Mater 2006, 16, 1299–1305.

64. Robert, F.S. Nanotechnology takes aim at cancer. Science 2005, 310, 1132–1134.

65. Delehanty, J.B.; Mattoussi, H.; Medintz, I.L. Delivering quantum dots into cells: Strategies, progress and remaining issues. Anal. Bioanal. Chem 2009, 393, 1091–1105.

66. Gao, X.H.; Cui, Y.Y.; Levenson, R.M.; Chung, L.W.; Nie, S.M. In vivo cancer targeting and imaging with semiconductor quantum dots. Nat. Biotechnol 2004, 22, 969–976.

67. Liu, J.; Stephen, K.L.; Vijay, A.V. Molecular mapping of tumor heterogeneity on clinical tissue specimens with multiplexed quantum dots. ACS Nano 2010, 4, 2755–2765.

68. Hardman, R. A toxicologic review of quantum dots: Toxicity depends on physicohemical and environmental factors. Environ. Health Perspect 2006, 114, 165–172.

69. Ciofani, G.; Raffa, V.; Menciass, A.; Cuschieri, A. Cytocompatibility, interactions, and uptake of polyethyleneimine-coated boron nitride nanotubes by living cells: Confirmation of their potential for biomedical applications. Biotechnol. Bioeng 2008, 101, 850–858.

70. Derfus, A.M.; Chan, W.C.; Bhatia, S.N. Intracellular delivery of quantum dots for live cell labeling of organelle tracking. Adv. Mater 2004, 16, 961–966.

71. Chen, H.; Wang, L.; Yeh, J.; Wu, X.; Cao, Z.; Wang, Y.A.; Zhang, M.; Yang, L.; Mao, H. Reducing non-specific binding and uptake of nanoparticles and improving cell targeting with an antifouling PEO-b-PgammaMPS copolymer coating. Biomaterials 2010, 31, 5397–5407.

72. Pellegrino, T.; Parak, W.J.; Boudreau, R.; le Gros, M.A.; Gerion, D.; Alivisatos, A.P.; Larabell, C.A. Quantum dot-based cell motility assay. Differentiation 2003, 71, 542–548.

73. Parak, W.J.; Boudreau, R.; le Gros, M.; Gerion, D.; Zanchet, D.; Micheel, C.M.; Williams, S.C.; Alivisatos, A.P.; Larabell, C. Cell motility and metastatic potential studies based on quantum dot imaging of phagokinetic tracks. Adv. Mater 2002, 14, 882–885.

74. Thein, M.; Cheng, A.; Khanna, P.; Zhang, C.; Park, E.J.; Ahmed, D.; Goodrich, C.J.; Asphahani, F.; Wu, F.; Smith, N.B.; Dong, C.; Jiang, X.; Zhang, M.; Xu, J. Site-specific sonoporation of human melanoma cells at the cellular level using high lateral-

resolution ultrasonic micro-transducer arrays. Biosens. Bioelectron 2011, 27, 25–33.

75. Susumu, K.; Uyeda, H.T.; Medintz, I.L.; Pons, T.; Delehanty, J.B.; Mattoussi, H. Design of biotin-functionalized luminescent quantum dots. J. Am. Chem. Soc 2007, 129, 13987–13996.

76. Delehanty, J.B.; Medintz, I.L.; Pons, T.; Brunel, F.M.; Dawson, P.E.; Mattoussi, H. Self-assembled quantum dot-peptide bioconjugates for selective intracellular delivery. Bioconjug. Chem 2006, 17, 920–927.

77. Rozenzhak, S.M.; Kadakia, M.P.; Caserta, T.M.; Westbrook, T.R.; Stone, M.O.; Naik, R.R. Cellular internalization and targeting of semiconductor quantum dots. Chem. Commun 2005, 17, 2217–2219.

78. Lieleg, O.; Lopez-Garcia, M.; Semmrich, C.; Auernheimer, J.; Kessler, H.; Bausch, A.R. Specific integrin labeling in living cells using functionalized nanocrystals. Small 2007, 3, 1560–1565.

79. Smith, B.R.; Cheng, Z.; De, A.; Koh, A.L.; Sinclair, R.; Gambhir, S.S. Real-time intravital imaging of RGD-quantum dot binding to luminal endothelium in mouse tumor neovasculature. Nano Lett 2008, 8, 2599–2606.

80. Pan, Y.L.; Cai, J.Y.; Qin, L.; Wang, H. Atomic force microscopy-based cell nanostructure for ligand-conjugated quantum dot endocytosis. Acta Biochim. Biophys. Sin 2006, 38, 646–652.

81. Yong, K.T.; Qian, J.; Roy, I.; Lee, H.H.; Bergey, E.J.; Tramposch, K.M.; He, S.; Swihart, M.T.; Maitra, A.; Prasad, P.N. Quantum rod bioconjugates as targeted probes for confocal and two-photon fluorescence imaging of cancer cells. Nano Lett 2007, 7, 761–765.

82. Zhang, H.; Sachdev, D.; Wang, C.; Hubel, A.; Gaillard-Kelly, M.; Yee, D. Detection and downregulation of type I IGF receptor expression by antibody-conjugated quantum dots in breast cancer cells. Breast Cancer Res. Treat 2009, 114, 277–285.

83. Lidke, D.S.; Nagy, P.; Heintzmann, R.; Arndt-Jovin, D.J.; Post, J.N.; Grecco, H.E.; Jares-Erijman, E.A.; Jovin, T.M. Quantum dot

ligands provide new insights into erbB/HER receptor-mediated signal transduction. Nat. Biotechnol 2004, 22, 198–203.

84. Liu, W.; Howarth, M.; Greytak, A.B.; Zheng, Y.; Nocera, D.G.; Ting, A.Y.; Bawendi, M.G. Compact biocompatible quantum dots functionalized for cellular imaging. J. Am. Chem. Soc 2008, 130, 1274–1284.

85. Diagaradjane, P.; Orenstein-Cardona, J.M.; Colón-Casasnovas, N.E.; Deorukhkar, A.; Shentu, S.; Kuno, N.; Schwartz, D.L.; Gelovani, J.G.; Krishnan, S. Imaging epidermal growth factor receptor expression in vivo: Pharmacokinetic and biodistribution characterization of a bioconjugated quantum dot nanoprobe. Clin. Cancer Res 2008, 14, 731–741.

86. Dudu, V.; Rotari, V.; Vazquez, M. Targeted extracellular nanoparticles enable intracellular detection of activated epidermal growth factor receptor in living brain cancer cells. Nanomedicine 2011, 7, 896–903. [Google Scholar]

87. Chakraborty, S.K.; Fitzpatrick, J.A.; Phillippi, J.A.; Andreko, S.; Waggoner, A.S.; Bruchez, M.P.; Ballou, B. Cholera toxin B conjugated quantum dots for live cell labeling. Nano Lett 2007, 7, 2618–2626.

88. Jaiswal, J.K.; Goldman, E.R.; Mattoussi, H.; Simon, S.M. Use of quantum dots for live cell imaging. Nat. Methods 2004, 1, 73–78.

89. Rajan, S.S.; Liu, H.Y.; Vu, T.Q. Ligand-bound quantum dot probes for studying the molecular scale dynamics of receptor endocytic trafficking in live cells. ACS Nano 2008, 2, 1153–1166.

90. Rajan, S.S.; Vu, T.Q. Quantum dots monitor TrkA receptor dynamics in the interior of neural PC12 cells. Nano Lett 2006, 6, 2049–2059. [Google Scholar]

91. Schroeder, J.E.; Shweky, I.; Shmeeda, H.; Banin, U.; Gabizon, A. Folate-mediated tumor cell uptake of quantum dots entrapped in lipid nanoparticles. J. Control. Release 2007, 124, 28–34.

92. Dudu, V.; Ramcharan, M.; Gilchrist, M.L.; Holland, E.C.; Vazquez, M. Liposome delivery of quantum dots to the cytosol of live cells. J. Nanosci. Nanotechnol 2008, 8, 2293–2300.

93. Voura, E.B.; Jaiswal, J.K.; Mattoussi, H.; Simon, S.M. Tracking metastatic tumor cell extravasation with quantum dot nanocrystals and fluorescence emission-scanning microscopy. Nat. Med 2004, 10, 993–998.

94. Duan, H.; Nie, S. Cell-penetrating quantum dots based on multivalent and endosome-disrupting surface coatings. J. Am. Chem. Soc 2007, 129, 3333–3338.

95. Fuente, J.M.; Fandel, M.; Berry, C.C.; Riehle, M.; Cronin, L.; Aitchison, G.; Curtis, A.S. Quantum dots protected with tiopronin: A new fluorescence system for cell-biology studies. ChemBioChem 2005, 6, 989–991. [Google Scholar]

96. Farias, PM.; Santos, B.S.; Menezes, F.D. Core-shell CdS/Cd(OH)2 quantum dots: Synthesis and bioconjugation to target red cells antigens. J. Microsc 2005, 219, 103–108.

97. Coulon, J.; Thouvenin, I.; Aldeek, F.; Balan, L.; Schneider, R. Glycosylated quantum dots for the selective labelling of Kluyveromyces bulgaricus and Saccharomyces cerevisiae yeast strains. J. Fluoresc 2010, 20, 591–597.

98. Bharali, D.J.; Lucey, D.W.; Jayakumar, H.; Pudavar, H.E.; Prasad, P.N. Folate-receptor-mediated delivery of InP quantum dots for bioimaging using confocal and two-photon microscopy. J. Am. Chem. Soc 2005, 127, 11364–11371.

99. Clarke, S.; Nadeau, J.; Bahcheli, D.; Zhang, Z.; Hollmann, C. Quantum dots as phototoxic drugs and sensors of specific metabolic processes in living cells. Proceedings of the 27th Annual International Conference of the Engineering in Medicine and Biology Society, Shanghai, China, 17–18 January 2005; Volume 1. pp. 504–507.

100. Xue, X.; Pan, J.; Xie, H.; Wang, J.; Zhang, S. Fluorescence detection of total count of Escherichia coli and

Staphylococcus aureus on water-soluble CdSe quantum dots coupled with bacteria. Talanta 2009, 77, 1808–1813.

101. Clarke, S.J.; Hollmann, C.A.; Zhang, Z.J.; Suffern, D.; Bradforth, S.E.; Dimitrijevic, N.M.; Minarik, W.G.; Nadeau, J.L. Photophysics of dopamine-modified quantum dots and effects on biological systems. Nat. Mater 2006, 5, 409–417.

102. Dubertret, B.; Skourides, P.; Norris, D.J.; Noireaux, V.; Brivanlou, A.H.; Libchaber, A. In vivo imaging of quantum dots encapsulated in phospholipid micelles. Science 2002, 298, 1759–1762.

103. Medintz, I.L.; Pons, T.; Delehanty, J.B.; Susumu, K.; Brunel, F.M.; Dawson, P.E.; Mattoussi, H. Intracellular delivery of quantum dot-protein cargos mediated by cell penetrating peptides. Bioconjug. Chem 2008, 19, 1785–1795.

104. Cai, W.B.; Shin, D.W.; Chen, K.; Gheysens, O.; Cao, Q.Z.; Wang, S.X. Peptide-labeled nearinfrared quantum dots for imaging tumor vasculature in living subjects. Nano Lett 2006, 6, 669–676.

105. Yu, X.; Chen, L.; Li, K.; Li, Y.; Xiao, S.; Luo, X. Immunofluorescence detection with quantum dot bioconjugates for hepatoma in vivo. J. Biomed. Opt 2007, 12, 1–5.

106. Weng, K.C.; Noble, C.O.; Papahadjopoulos-Sterberg, B.; Chen, F.F.; Drummond, D.C.; Kirpotin, D.B. Targeted tumor cell internalization and imaging of multifunctional quantum dot-conjugated immunoliposomes in vitro and in vivo. Nano Lett 2008, 8, 2851–2857.

107. Janib, S.M.; Moses, A.S.; MacKay, J.A. Imaging and drug delivery using theranostic nanoparticles. Adv. Drug Deliv. Rev 2010, 62, 1052–1063.

108. Matsuno, A.; Mizutani, A.; Okinaga, H.; Takano, K.; Yamada, S.; Yamada, S.M.; Nakaguchi, H.; Hoya, K.; Murakami, M.; Takeuchi, M.; Sugaya, M.; Itoh, J.; Takekoshi, S.; Osamura, R.Y. Molecular morphology of pituitary cells, from conventional

immunohistochemistry to fluorescein imaging. Molecules 2011, 16, 3618–3635.

109. Byers, R.J.; Hitchman, E.R. Quantum dots brighten biological imaging. Prog. Histochem. Cytochem 2011, 45, 201–237.

110. Isse, K.; Grama, K.; Abbott, I.M.; Lesniak, A.; Lunz, J.G.; Lee, W.M.; Specht, S.; Corbitt, N.; Mizuguchi, Y.; Roysam, B.; Demetris, A.J. Adding value to liver (and allograft) biopsy evaluation using a combination of multiplex quantum dot immunostaining, high-resolution whole-slide digital imaging, and automated image analysis. Clin. Liver Dis 2010, 14, 669–685.

111. Tholouli, E.; Sweeney, E.; Barrow, E.; Clay, V.; Hoyland, J.A.; Byers, R.J. Quantum dots light up pathology. J. Pathol 2008, 216, 275–285.

112. Smith, A.M.; Nie, S. Chemical analysis and cellular imaging with quantum dots. Analyst 2004, 129, 672–677.

113. Yaghini, E.; Seifalian, A.M.; MacRobert, A.J. Quantum dots and their potential biomedical applications in photosensitization for photodynamic therapy. Nanomedicine 2009, 4, 353–363.

114. Rakovich, A.; Savateeva, D.; Rakovich, T.; Donegan, J.F.; Rakovich, Y.P.; Kelly, V.; Lesnyak, V.; Eychmüller, A. CdTe quantum dot/dye hybrid system as photosensitizer for photodynamic therapy. Nanoscale Res. Lett 2010, 5, 753–760.

8

CHAPTER

PHYSICAL REASONS OF EMISSION VARYING IN CDSE/ZNS AND CDSETE/ZNS QUANTUM DOTS AT BIOCONJUGATION TO ANTIBODIES

Tetyana V. Torchynska[1]

[1] *National Polytechnic Institute, ESFM, Mexico D.F., Mexico*

1. INTRODUCTION

In the last two decades, the colloidal core/shell quantum dots (QDs) of II-VI semiconductors (CdSe, CdS, CdSeTe, and ZnS) attracted an enormous scientific attention owing to the fundamental scientific aspects and future promising applications in optoelectronics, photonics, biology, and medicine. The modern developments in nanotechnology allow numerous promising applications in the field of biology and medicine, which in the future may bring an unprecedented effect in nanomedicine [1, 2].

The development of biocompatible nanoparticles for molecular imaging and targeted therapy is an area of enormous current interest. This novel technique as expected will allow targeted drug delivery to cells with certain parameters, as well as improvements in diagnostics and localization of affected tissues in the human body.

When conjugated with biomolecular affinity ligands, such as antibodies, peptides, or small molecules, these nanoparticles can be used to target malignant specific tumors [3-5].

The key issue for nanoparticles intended for biological use is to ensure their stability, which suggests the formation of core-shell structures with a certain enveloping layer required to isolate a nanoparticle [6]. The core defines the main properties of the system—it may be the material with outstanding luminescence or magnetic properties for the precise detection of particle's location inside the body. On the other hand, the core can be used as a reservoir for the medicine used in targeted drug delivery applications. The application of semiconductor QDs as luminescent markers in biology and medicine is expected to produce the major advances in molecular diagnostic [1], gene technology [2], toxin detections [7], drag delivery [8], at obtaining tissue imaging in vivo etc [3, 9].

Note that luminescent markers usually used earlier, such as organic dyes, fluorescent proteins, or lanthanide chelates, have some restrictions, such as broad spectrum bands, low photo bleaching threshold, poor photochemical stability, and degradation [10]. In contrary, the II-VI semiconductor core/shell QDs are characterized by the high-photoluminescence (PL) quantum yield (up to 75%) [11-13]. Surface-passivized II-VI QDs are highly stable against photo bleaching and are characterized by narrow, symmetric emission peaks with the half width about 25-30 nm [6].

New luminescent markers are needed for the better assessment of treatments for many types of cancers. The antibodies, which can be found in circulating blood, may help in the early detection of cancer. The development of a blood test, using the optical methods, which permits early diagnosis, would be a great advance in the management of cancer disease. Note that the core/shell CdSeTe/ZnS QDs with near-infrared (IR) emission (780-800 nm) are very interesting for biological applications owing to the possibility of animal imaging in vivo due to the low absorption of IR light by an animal tissue [6, 14, 15]. At the same time, the chemical behavior of Cd/Se/Te components of QDs in the biological system is still unclear.

The last question is very important due to the toxicity of these components for the human body [16-18].

The conjugation of biomolecules with semiconductor core/shell QDs has been achieved using functional groups (linkers) on the QD surface [19-21], and/or with the help of electrostatic interaction between the QDs and the biomolecules [12, 22, 23]. A set of publications, related to the study of QD bioconjugation using PL spectroscopy, revealed that the emission intensity of QDs decreased [12, 24, 25] or increased [22, 26] owing to the energy exchange between the QDs and the biomolecules. Thus, the QD luminescence intensity depends on the concentration of attached biomolecules, promising the QD application as a protein sensor as well [6]. However, the limits of biomolecule detection are not very sensitive. Moreover, the influence of bioconjugation processes on QD optical properties and the bioconjugation mechanisms are not yet completely understood.

It is desirable to have the additional information concerning the conjugation process such as a spectral shift of QD emission or changing the PL band half width, or using nonlinear optical phenomena: two-photon-induced fluorescence, second harmonic generation, or sum frequency generation [27]. Simultaneously, the other optical methods can be useful for the bioconjugation detection such as the Raman scattering method [28] or coherent anti-Stokes Raman scattering (CARS) [27]. This chapter presents the study of the transformation of PL and Raman scattering spectra of the core/shell CdSe/ZnS QDs with visible emission (605 or 655 nm) and CdSeTe/ZnS QDs with near IR emission (780-800 nm) at the bioconjugation to different types of antibodies.

2. QD BIOCONJUGATION PROCESS AND EXPERIMENTAL DETAILS

Core-shell quantum dots (QDs) commercially available covered by the amine (NH_2)-derivatized polyethylene glycol (PEG) polymer were used in a form of colloidal particles diluted in a phosphate-buffered saline (PBS) with a 1:200 volumetric ratio. CdSe/ZnS QDs with emission peaked at 605 nm (2.05 eV) and 655 nm (1.89 eV) and CdSeTe/ZnS QDs with emission at 780-800 nm (1.55-1.60 eV) have been investigated. QDs have been studied by means of photoluminescence and Raman scattering methods in nonconjugated states and after the conjugation to different types of antibodies (Ab). All optical measurements are performed on the dried droplets of the original solution of nonconjugated and bioconjugated QDs. At the first the PL spectra of QDs are studied in the nonconjugated state. Then some part of QDs has been conjugated to the antibodies using the QD conjugation kits [29, 30]. These kits contain amine-derivatized polymer-coated QDs and the amine-thiol cross-linker SMCC. The samples of QDs (bioconjugated and nonconjugated) in the form of a 5-mm spot were dried on a surface of crystalline Si substrates (Figure 1) as described earlier by Torchynska [22] and Vega Macotela et al. [24].

The types of antibodies (Ab) used for the bioconjugation to the CdSe/ZnS QDs with emission at 605 nm (2.05 eV) and 655 nm (1.89 eV) are as follows:

i. Mouse monoclonal [8C9] antihuman papilloma virus Ab, HPV 16-E7, bacterially derived fusion protein containing anti-HPV 16 early protein E7 Ab (Invitrogen, isotype (IgG1))

ii. Anti-interleukin-10 (IL-10) antibodies (antihuman IL-10, Rt IgG1, stock concentration of 1 mg/ml, clone JES3-9D7, code RHCIL1000)

iii. Pseudorabies virus (PRV) Ab, immunoglobulin G antibodies (affinity purified with Protein G Sepharose from rabbit antiserum to pseudorabies virus, stock concentration of 1 mg/ml in PBS)

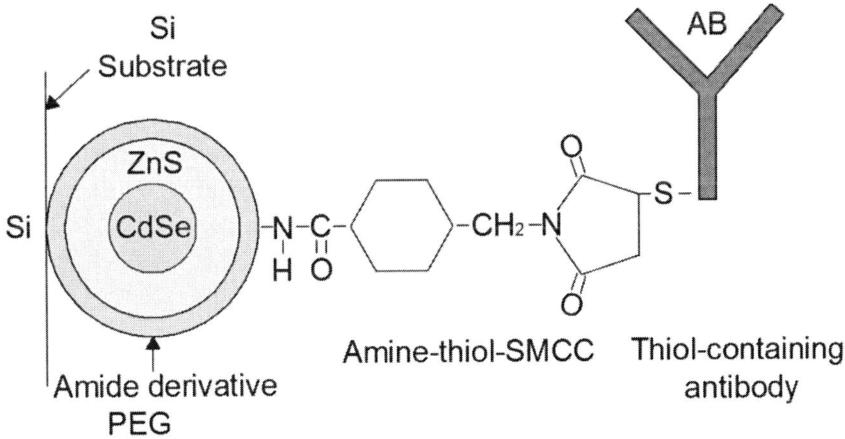

Figure 1. Bioconjugated CdSe/ZnS QDs on the Si substrate.

CdSeTe/ZnS QDs with near IR emission at 780-800 nm (1.59-1.60 eV) have been conjugated to different types of antibodies as well:

i. Mouse monoclonal [8C9] antihuman papilloma virus Ab, anti-HPV 16-E7, bacterially derived fusion protein containing anti-HPV 16 early protein E7 Ab (Invitrogen, isotype (IgG1))

ii. Mouse monoclonal [C1P5] antihuman papilloma virus HPV16 E6 + HPV18 E6 Ab, HPVE6 (Abcam, ab70, isotypeIgG1, gel-purified HPV18 E6-betagalacto-sidase fusion protein)

iii. Pseudorabies virus (PRV) Ab, immunoglobulin G antibodies (affinity purified with Protein G Sepharose from rabbit antiserum to Pseudorabies virus, stock concentration of 1 mg/ml in PBS)

iv. Anti-IL-10 antibodies (antihuman IL-10, Rt. IgG1, stock concentration of 1 mg/ml, clone JES3-9D7, code RHCIL1000)

The protocol of bioconjugation details can be found in refs. [29, 30]. Bioconjugated CdSe/ZnS QDs have been called as 605P+IL-10, 655P+IL-10, 605P +HPV E7, 655P +HPV E7, 605P+PRV, and

655P+PRV, and the bioconjugated CdSeTe/ZnS QDs have been called as (i) 800P+HPV E7, (ii) 800P+HPV E6, (iii) 800P+PRV, and (iv) 800P+IL-10.

The majority of PL spectra were measured at the excitation by a He-Cd laser with a wavelength of 325 nm and a beam power of 76 mW at 300 K using a PL setup on a base of spectrometer SPEX500 described by Torchynska [22] and Vega Macotela et al. [24]. Raman scattering spectra were measured at 300 K by means of the spectrometer Lab-Raman HR800 Horiba Jovin-Yvon in the range of Raman shifts of 100-600 cm^{-1} at the excitation by a solid-state laser with a wavelength of 532 nm and a beam power of 20 mW [26, 28].

3. THE BIOCONJUGATION STUDY OF CDSE/ZNS QDS

3.1. CdSe/ZnS QDs bioconjugated to anti-IL-10 Ab

Figures 2 and 3 demonstrate the PL spectra of the three samples of nonconjugated (605N, 655N) QDs and three samples of bioconjugated (605P, 655P) QDs.

The PL spectra of CdSe/ZnS QDs are characterized by the Gaussian shape PL band in the nonconjugated states with the peaks at 2.04 eV (Figure 2) and 1.90 eV (Figure 3) related to exciton emission in corresponding CdSe cores [31]. At the bioconjugation of studied QDs, the PL intensity decreases, the emission peak shifts to higher energies, a PL half width increases, and the shape of PL bands became asymmetric with high-energy tails (Figures 2 and 3). The PL energy shift at the bioconjugation of QDs to antibodies can be attributed to the following: (i) the impact of anti-IL-10 Ab or PBS buffer emission excited by UV light, (ii) the oxidation or degradation of CdSe cores owing to core/shell atom intermixing [32], (iii) the impact of compressive strains that appears at the bioconjugation [33, 34], or (iv) the emission of excitons bound to the QD excited states in QDs [31]. To distinguish between proposed reasons, the PL spectra of anti-IL-10 Ab and PBS were studied at UV light excitation (Figure 4).

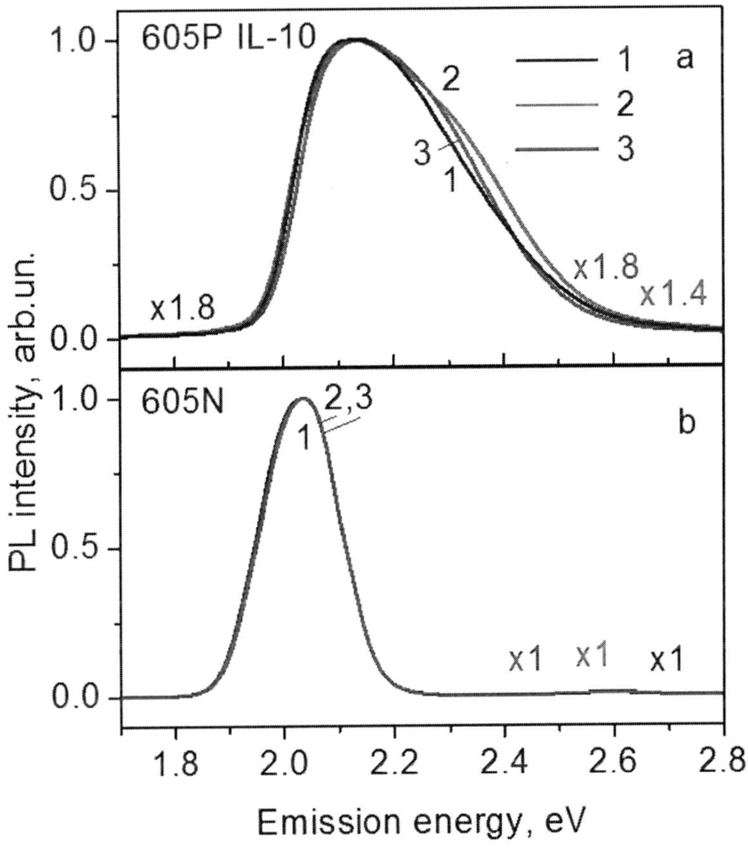

Figure 2. Normalized PL spectra of three 605P (a) and three 605N (b) QD samples at 300 K. Numbers at the curves (×1, ×1.4, and ×1.8) show the multiplication coefficients used at the normalization of PL spectra.

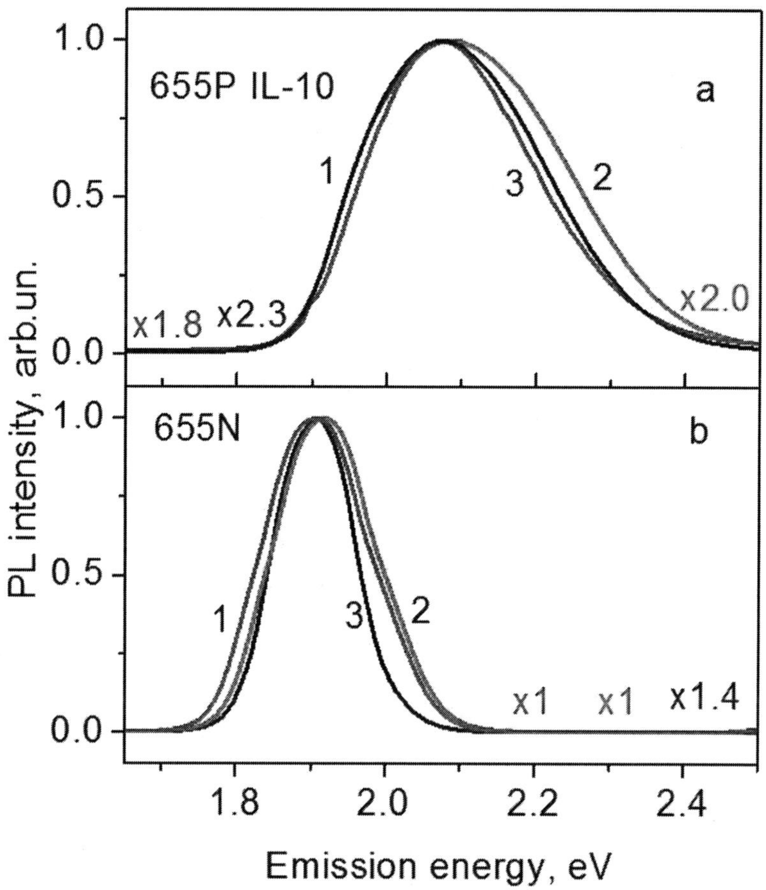

Figure 3. Normalized PL spectra of three655P (a) and three 655N (b) QD samples at 300 K. Numbers at the curves (×1, ×1.8, and ×2.0) show the multiplication coefficients used at the normalization of PL spectra.

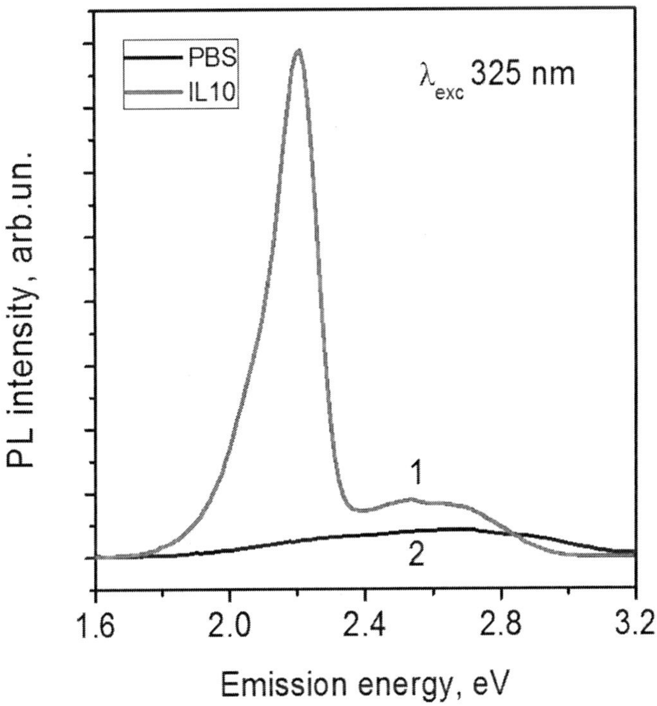

Figure 4. PL spectra of anti-IL-10 Ab (1) and PBS (2) without QDs.

The PL spectrum of anti-IL-10 Ab is characterized by the PL band with a maximum at 2.21 eV and a small shoulder in the range 2.4-3.0 eV (Figure 4, curve 1). The PL spectrum of PBS is shown in Figure 4, curve 2. The PL intensity of PBS in the range 2.2-3.2 eV is 15-fold smaller than the emission intensity of 2.21 eV PL band related to anti-IL-10 Ab. Simultaneously, the PL intensity of anti-IL-10 Ab emission (2.21 eV) is 10-fold smaller than the PL band intensity of CdSe/ZnS QDs at UV excitation. Hence, we can conclude that the "blue" shifts of PL spectra in bioconjugated QDs (605P and 655P) do not connect with the luminescence of anti-Il-10 Ab or PBS. However, the emission of anti-IL-10 Ab cans influent on the shape of tails in PL spectra of bioconjugated QDs.

Raman scattering spectra were investigated with the goal to study the impact of compressive strains [33, 34] and the oxidation or degradation processes at the QD bioconjugation [32, 35], as well as to confirm the existence of electric charges in IL-10 Ab. Raman scattering spectra of 605N and 605P CdSe/ZnS QDs present the low-intensity Raman peaks at 211.9, 235.8, 302.8, 345.3, 424.8, 434.6, and 490 cm^{-1} (Figure 5).

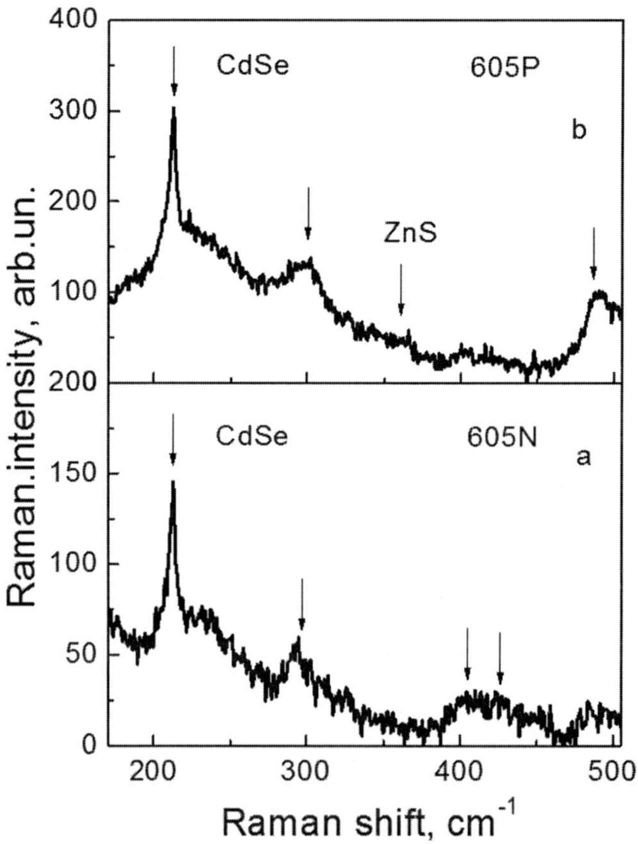

Figure 5. Raman scattering spectra of nonconjugated (a) and bioconjugated (b) QDs.

The peak at 211.9 cm^{-1} (and two phonon overtone at 424.8 cm^{-1}) corresponds to LO phonons in the CdSe core. The shift of LO phonon Raman peak (211.9 cm^{-1}) from its position in the bulk CdSe crystal (213 cm^{-1}) has related to the phonon confinement effect in small QDs [6, 35]. The peak at 345.3 cm^{-1} is due to the LO phonon scattering in the ZnS shell (Figure 5). The shift of LO phonon Raman peak (345.3 cm^{-1}) from its position in the bulk ZnS crystal (352 cm^{-1}) is related to the phonon confinement effect in thin ZnS shell [6, 35, 36]. Raman spectra in the diapason of 0-500 cm^{-1} include the peaks related to the Si substrate as well (Figure 1). Raman scattering spectra of Si present the overtones of acoustic phonons: 235.8, 302.8, 434.6, and 458.6 cm^{-1} (Figure 5). The peaks at 230, 302, 435, and 469 cm^{-1} were attributed early to two TA phonon overtones at the L, X, and Σ critical points of the Si Brillouin zone [36-38].

The Raman study shows that Raman peaks related to LO phonons in the CdSe core (Figure 5) have not varied at the QD bioconjugation. Meanwhile, if the QD bioconjugation is accompanied by the compressive strain applied, and by the degradation or oxidation of cores, the Raman peaks have to change. Finally, we can conclude that the mentioned processes have not been realized at the QD bioconjugation to anti-IL-10 Ab.

Note that the CdSe core Raman peak intensity increases twofold in bioconjugated QDs as it is shown in Figure 5. The enhancement of optical field at the surface of bioconjugated QDs and increasing the Raman peak intensity can be attributed to the surface-enhanced Raman scattering (SERS) [28,39]. This fact means that anti-IL-10 Abs have the electric charges (dipoles), which interact with an electric field of excitation light at the SERS effect [39-41].

To study the impact of excitons at QD excited states on the emission of bioconjugated QDs, the PL spectra were measured at different UV light (325nm) excitation intensities. The PL spectra of 655N QDs kept a Gaussian shape of PL band (1.90 eV) at all excitation intensities (Figure 6a). The full width at half maximum (FWHM) of PL band at the first increases with raising excitation intensities owing to the enlargement of the number of excited QDs and their size distribution. Then the FWHM saturates at higher excitation intensities (Figure 6c, curve 1).

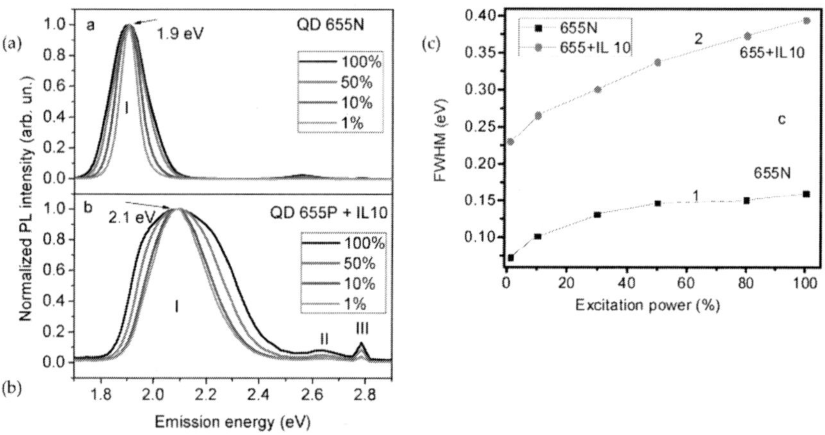

Figure 6. (a,b). Normalized PL spectra of 655N (a) and 655P-IL-10 (b) QDs at different excitation light intensities 100%, 50%, 10%, and 1%, respectively. The 100% intensity corresponds to 76 mW. Coefficients at the normalization of PL spectra for 655N QDs (a) are ×1.00, ×1.10, ×1.75, and ×9.11 and for 655P QDs (b) are ×1.28, ×1.59, ×3.38, and ×31.18. (c). FWHMs of main PL bands in PL spectra of 655N (1) and 655P-IL-10 (2) QDs versus excitation light power.

The PL spectra of bioconjugated QDs represent a set of PL bands: (i) PL band (I) with a peak at 2.10 eV connected with QD core emission; (ii) PL band (II) at 2.64 eV, apparently, is related to PBS emission; and (iii) PL band (III) with the peak at 2.78 eV (Figure 6b).

The high-energy PL band III is attributed to a CdSeS alloy appeared at the CdSe/ZnS interface [31]. The confirmation of the existence of such alloy has been obtained at the Raman scattering study in CdSe/ZnS QDs (see Figure 5). The formation of high-energy tails of PL band I together with increasing the PL band II and III intensities have been detected by raising the excitation light power (Figure 6b). Simultaneously, the intensity of a low-energy part of band I increases versus excitation power due to the increasing excited QD numbers and QD size distribution, as it has been seen in nonconjugated QDs. The FWHM of band I in bioconjugated QDs increases monotonously without saturation versus excitation intensity (Figure 6c, curve 2). This study has shown that the PL band I intensity increases with excitation power owing to the increasing emission of excitons localized at the excited states in bioconjugated QDs. The mentioned effect can explain the asymmetric shape of band I and the increasing FWHM (Figure 6c, curve 2) in bioconjugated QDs. Note that the exciton emission via excited states cannot explain "blue" shift (~200 meV) in the PL band I peak and appearing the PL band III in the PL spectra of QDs bioconjugated to anti-IL-10 Abs.

3.2. CdSe/ZnS QDs bioconjugated to HPV E7 and PRV Ab

The PL spectra of CdSe/ZnS QDs with emission 605 nm and 655 nm nonconjugated and bioconjugated to PRV and HPV E7 antibodies have been presented in Figures 7 and 8, respectively. The PL spectra of QDs in a nonconjugated state are characterized by the one Gaussian shape PL band peaked at 2.05 eV (Figure 7a) and 1.90 eV (Figures 7b and 8a) and related to exciton emission in corresponding CdSe cores. The PL spectra of bioconjugated QDs have changed essentially: the core PL band shifts into the high-energy spectral range ("blue" shift). Simultaneously, the PL intensity of core emission decreases at the bioconjugation, its half width increases, and the shape of PL band becomes asymmetric with the essential high-energy tails (Figures 7 and 8).

As we have shown above, the processes of oxidation, the CdSe material degradation, or the appearance of elastic strains have been not realized at the bioconjugation of QDs [31, 42]. The "blue" energy shift of PL bands in bioconjugated QDs can be assigned to the emission of HPV E7 or PRV antibodies or PBS at high-energy UV excitation and/or the emission of excitons localized at the excited QD states.

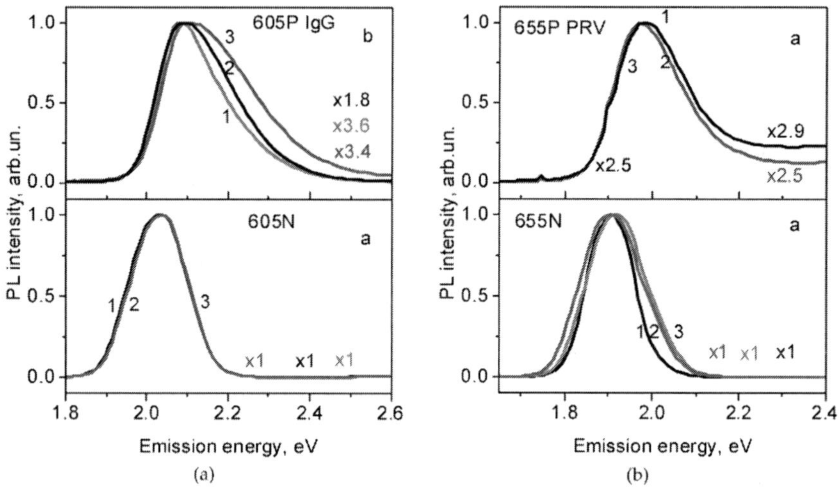

Figure 7. (a). Normalized PL spectra of three different 605N (a) and 605P-PRV (b) QD samples measured at 300 K. Numbers at the curves (×1, ×1.8, ×3.4, and ×3.6) indicate on the multiplication coefficients used at the normalization. (b). Normalized PL spectra of three different 655N (a) and 655P-PRV (b) QD samples measured at 300 K. Numbers at the curves (×1, ×2.5, and ×2.9) indicate on the multiplication coefficients used at the normalization.

The PL spectra of HPV E7 Ab, PRV Ab, and PBS have been studied at UV excitation as well (Figure 9). As it is clear from Figure 9, the PL spectra of antibodies and PBS are characterized by PL bands in the spectral range of 2.0-3.2 eV. The intensity of PL band in the spectral range 2.0-3.2 eV related to PBS is five- or eightfold smaller than the intensity of PL band related to antibodies. Simultaneously, the intensity of PL band related to the antibodies 10-fold smaller than the PL intensity of CdSe/ZnS QD emission at the UV excitation. Thus, we can conclude as well that the "blue" energy shift of PL spectra in bioconjugated QDs does not connect with the emission of HPV E7 Ab, PRV Ab or PBS.

To study the role in the emission of excitons, localized at QD excited states in bioconjugated QDs, the PL spectra were measured at different excitation light intensities (Figure 8a, b). The PL spectra of nonconjugated 655N QDs kept the Gaussian shape of core PL band (1.90 eV) for all excitation intensities used (Figure 8a). The FWHM of this PL band at the first increases versus excitation intensity, due to raising the excited QD numbers and QD size distribution, and then the FWHM saturates at higher excitation intensities (Figure 8c, curve 1).

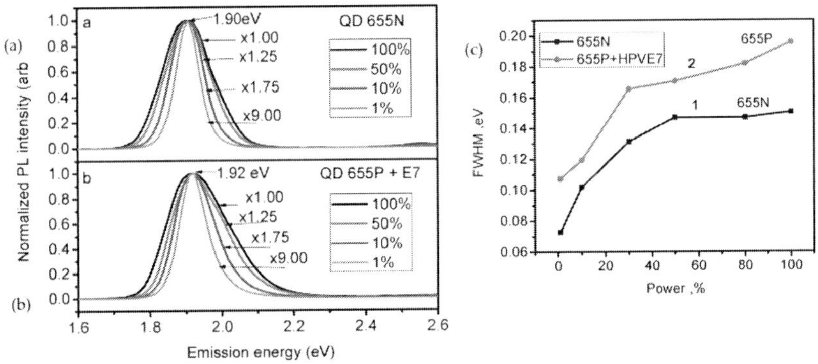

Figure 8. (a,b). Normalized PL spectra of 655N (a) and 655P-HPV E7 (b) QD samples measured at 300 K at the excitation intensities 100%, 50%, 10%, and 1%. The excitation intensity of 100% corresponds to 76 mW. (c). The FWHM variation of PL bands in 655N (1) and 655P-HPV E7 (2) QDs versus excitation light power. 100% corresponds to the power 76 mW.

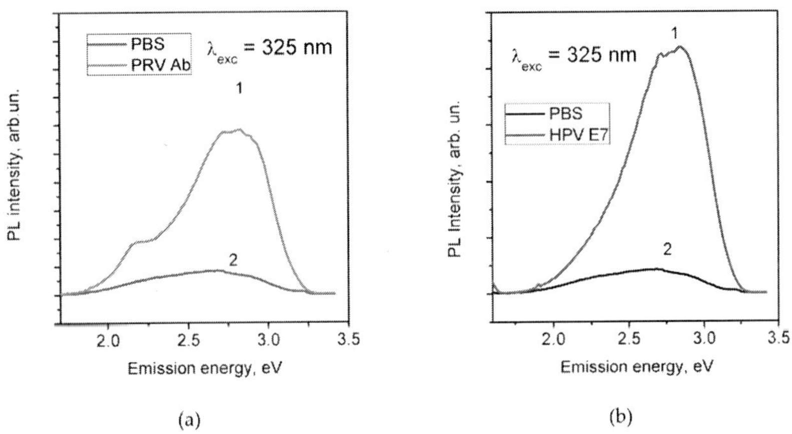

Figure 9. (a)PL spectra of PRV Ab (1) and PBS (2) without QDs. (b) PL spectra of HPV E7 Ab (1) and PBS (2) without QDs.

The PL band intensity in QDs, bioconjugated to HPV E7 Abs, increases versus excitation light power together with the formation of high-energy tails (Figure 8b). The PL band intensity at a low-energy side increases owing to rising the excited QD numbers and QD size distribution by the same way as in nonconjugated QDs. The FWHM of PL band in bioconjugated QDs increases monotonously versus excitation intensity without the saturation effect (Figure 8c, curve 2). Thus, the PL band intensity at a high-energy side increases versus excitation light power due to raising the role of exciton emission via excited states in bioconjugated QDs. This effect can explain an asymmetric shape of PL bands and increasing the FWHMs (Figure 8c, curve 2) in bioconjugated QDs. However, the emission of excitons localized at the excited QD states cannot explain shifting the main PL peak to higher energy in the PL spectra of bioconjugated CdSe/ZnS QDs.

4. DISCUSSION OF THE CDSE/ZNS QD BIOCONJUGATION

The experimental results presented above have shown that the influence of exciton emission via excited QD states or the emission of antibodies and PBS as well as the oxidation or degradation processes in QD cores can be avoided from the consideration as the reasons of "blue" PL peak shift at the QD bioconjugation. Let us discuss other reasons for the PL energy shift at the bioconjugation of QDs to antibodies. These factors could be assigned to (i) the quantum-confined Stark effect stimulated by the charge of antibodies [43, 44], (ii) the quantum-confined effect related to the shift of QD energy levels and stimulated by the change of electric potential at the QD surface [45], or (iii) decreasing the effective QD size in bioconjugated QDs [46].

The Stark effect is shifting and splitting of energy levels in atoms and molecules due to the presence of external electric field. The Stark shift of energy levels, ΔE, as a function of electric field, ξ, can be presented as a sum of linear (first order Shark effect) and quadratic (second order Stark effect) functions of an electric field: $\Delta E = \mu_{QD}\,\xi + 0.5\,\alpha_{QD}\,\xi^2 + ...$, where E is an energy of optical transition, and μ_{QD} and α_{QD} are projections of excited-state dipole and polarizability, respectively, along the electric field [43, 44]. This relation includes both the polar and polarizable parts of emitting state. Actually, the Stark shift in the emission of QD ensembles was found earlier to be purely quadratic versus applied electric field [39]. The dependences of the quantum-confined Stark shift versus QD sizes and electric fields in the CdS, CdSSe, and CdSe QDs were studied theoretically [44] and experimentally [43] in the early 90th. Only the "red" Stark energy shift was predicted theoretically for CdS and CdSSe QD ensembles at applying an external electric field [44]. Thus, we can conclude that the quantum-confined Stark effect could not explain the observed experimental "blue" energy shift of PL bands at the QD bioconjugation to studied antibodies.

Another reason for the "blue" shift of PL peak in bioconjugated QDs can be connected with the change of potential barrier at the surface of QDs bioconjugated to the charged antibodies. The position of energy levels in QDs for the strong quantum-confined regime depends on the value of potential barrier [45]. It was shown that the impact of a finite potential barrier on the electron-hole energy states increases with decreasing the dot radius. The lower potential barrier reduces and higher potential barrier increases the energy levels and these changes are more relevant for high-energy states [45]. However, the comparison of the PL spectrum transformation presented in Figures 2 and 3 or Figure 7a, b has shown that the value of "blue" PL energy shift at the QD bioconjugation to anti-IL-10 or PRV Abs is more essential for QDs with the biggest sizes (6.4 nm) and emission at 655 nm. Thus, the "blue" PL band energy shifts at the CdSe/ZnS QD bioconjugation to antibodies do not connect with the variation of the potential barrier at the QD surface.

The other reason for the "blue" shift of QD energy levels at the quantum-confined condition, owing to decreasing the effective QD size in bioconjugated QDs, was considered by Torchynska et al. [46]. When QD is bioconjugated, its energy balance changes very much. The energy variation comes from the fact that the attachment of biomolecules to the external QD surface will result in the generation of van der Waals forces. The extra charge attached to the exterior shell will form a blocking electric field, which "truncates" the effective volume of QDs (Figure 10). In this situation, it is natural to expect energy varying of emitted light. At the same time, the stronger confinement will produce a blue shift of the QD emission spectrum.

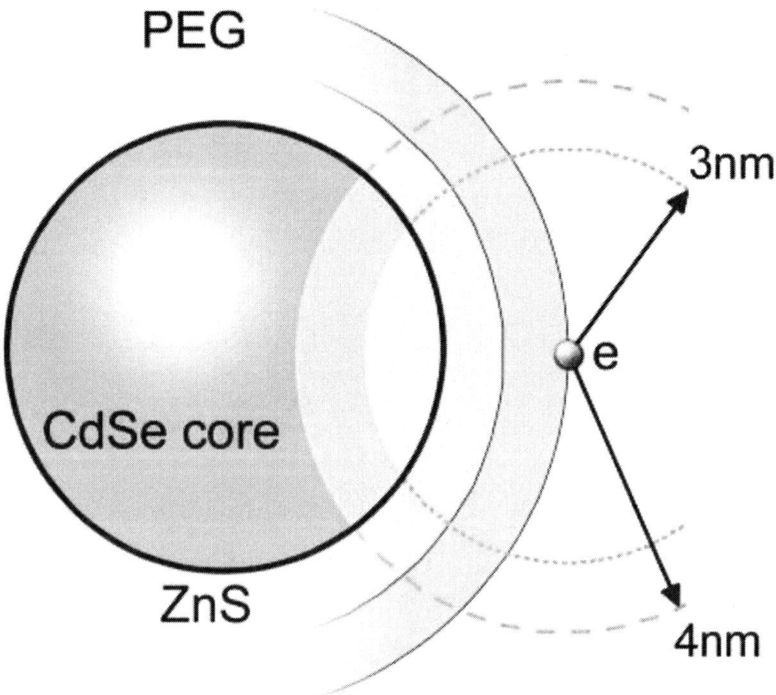

Figure 10. A schematic depiction of CdSe/ZnS QD with the 6.4 nm CdSe core, encased in the ZnS and poly-ethylene glycol (PEG) protective shells. A charge generated upon bioconjugation is shown attached to the PEG shell to the right. We consider two radii of the blocking sphere of $d = 3$ and 4 nm, respectively.

The question about the accurate calculation of the truncated volume is a complicated one. For a point charge, the equipotential surfaces will be spheres, which will make the blocked volume corresponding to a spherical segment. The remaining effective volume will lose its spherical shape. We proposed to approximate it with a cylinder with a height c and a circular base of the diameter a. The Schrödinger equation for such system was solved using the mirror boundary conditions [46, 47], providing the ground state energy relative to the material band gap [47]:

$$E_{GSc} - E_g = \frac{h^2}{2m}\left(\frac{0.268}{a^2} + \frac{0.25}{c^2}\right) \tag{1}$$

The parameters of a cylinder can be estimated for different strength of the blocking field. We consider two cases with the blocking field extending for the distance d = 3 and 4 nm from the charge attached to the protective PEG layer covering nanoparticle (Figure 9). For these cases, the parameters of the effective cylinder are as follows: a = 1.6 R, c = 1.55 R (for 4 nm blocking radius) and a = 1.65 R, c = 1.6 R (for 3 nm blocking radius), where R = 3.2 nm is the radius of the CdSe core of the particle. Using these values, one can obtain the expressions for the ground state of conjugated nanoparticle with 4 nm blocking [46]:

$$E_{GS_c} - E_g \approx \frac{h^2}{9.6mR^2} \tag{2}$$

and also for conjugated nanoparticle with 3 nm blocking distance [46]:

$$E_{GSc} - E_g \approx \frac{h^2}{11.1mR^2} \tag{3}$$

In a real system, the absorbed energy that triggered the formation of an exciton can be partially dissipated, so that the peak of PL spectra will occur at lower energy, producing Stokes shift E_{St}. Therefore, the total blue shift observed in the system can be defined as a ratio between peak positions of a bioconjugated sample $h\nu_c$ versus nonconjugated sample $h\nu_0$:

$$\frac{E_{GS_c} - E_g}{E_{GS_0} - E_g} = \frac{h\nu_c + E_{St} - E_g}{h\nu_0 + E_{St} - E_g} \tag{4}$$

Using the experimental data shown in Figure 3, one can see that $h\nu_0$ = 1.89 eV and $h\nu_c$ = 2.08 eV; the typical Stokes shift is E_{St} = 56.0 meV [48]. Using the band gap E_g = 1.74 eV for the bulk CdSe at 300 K [49], one can obtain the ratio for the experimental shift equal to 2.16 from equation (4). The ratios calculated with Eqs. (1), (2), (3), and (4) are 3.30 (for blocking distance d = 4 nm) and 2.88 (for d = 3 nm), which is considerably close to the experimental value in view of simplifications made in treating the truncation of nanoparticle's core volume.

Thus, we have proposed to use the mirror boundary condition for simplifying solution of the Schrödinger equation for a spherical CdSe/ZnS core/shell QDs. It was shown that the "blue" shift in the PL spectrum observed in the experiment with bioconjugated QDs can be explained by the attachment of a charge deals with the antibody to the exterior shell of QDs, blocking away a fraction of core's volume. Representing the truncated effective part of core as a cylinder, we obtained a blue shift values with a considerable accuracy in relation to the experimental data. Thus, the "blue" energy shift of PL spectrum in the bioconjugated CdSe/ZnS QDs has been attributed to the electronic quantum-confined effects stimulated by the charged antibodies.

5. THE BIOCONJUGATION STUDY OF CDSETE/ZNS QDS

5.1. CdSeTe/ZnS QDs bioconjugated to anti-IL-10 Ab

The PL spectra of CdSeTe/ZnS QDs for the nonconjugated (800N) and bioconjugated (800P+ IL-10) states are shown in Figure 11. The PL spectrum of 800N QDs is characterized by Gaussian shape PL band with a peak at 1.59-1.60 eV (Figure 11, curves 1 and 2) connected with exciton emission in CdSeTe cores. The PL peak shifts to higher energies ("blue" shift) up to 1.90 eV (Figure 11, curves 3 and 4), and the shape of PL bands becomes asymmetric with low-energy shoulders (at 1.75 eV) in the PL spectra of bioconjugated QDs (Figure 11, curves 3 and 4). Simultaneously, the PL intensity of QDs decreases in the case of the bioconjugation to anti-IL-10 Ab (Figure 11, curves 3 and 4).

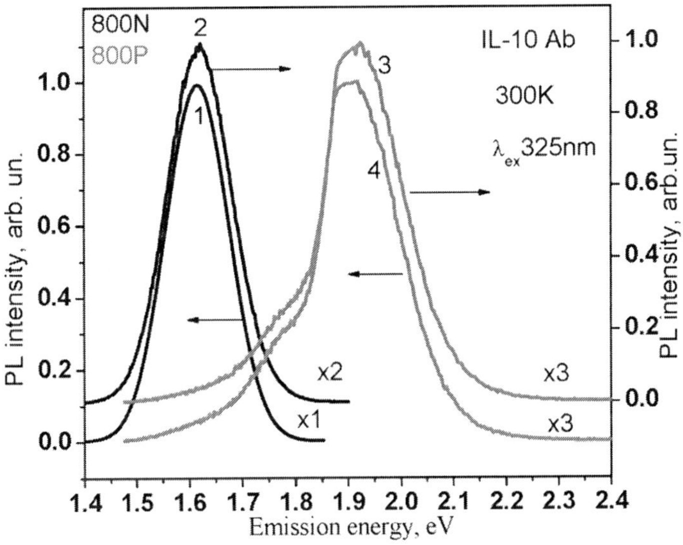

Figure 11. Normalized PL spectra of two 800N (curves 1 and 2) and two 800P+ IL-10 (curves 3 and 4) QD samples at 300 K. PL spectra are shifted along *y* axis to prevent their overlapping. Numbers (x1, x2, x3) indicate the multiplication coefficients used at the normalization. Excitation light wavelength is 325 nm.

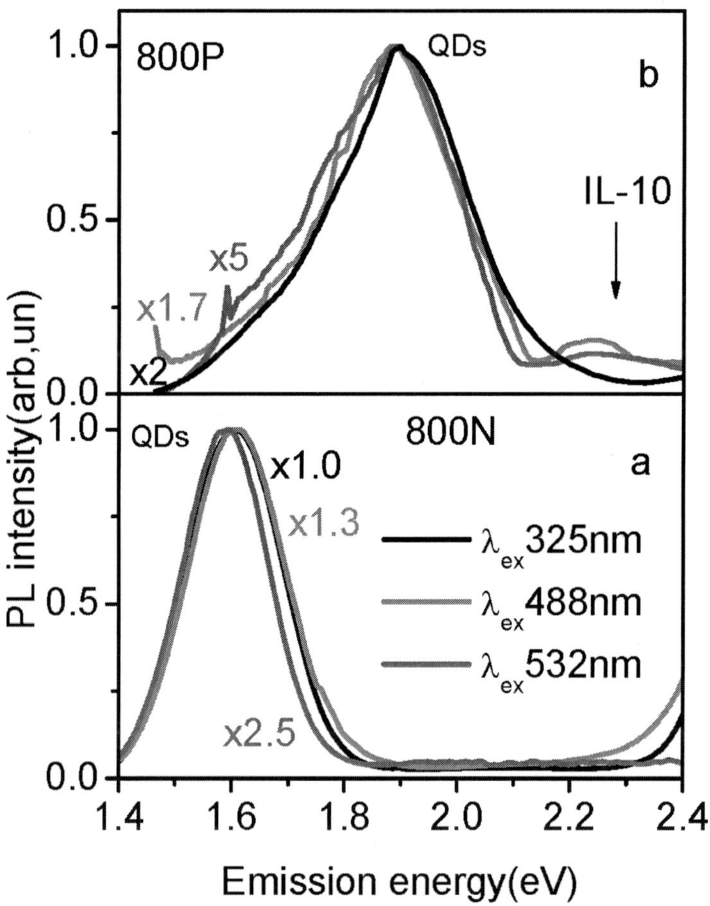

Figure 12. PL spectra of 800N (a) and 800P+ IL-10 (b) CdSeTe/ZnS QDs measured at 300 K and excited by the light with wavelengths: 532 nm (1), 488 nm (2), and 325 nm (3).

The PL spectra of anti-IL-10 Ab and PBS have been measured at the UV light excitation early (Figure 4). The PL spectrum of anti-IL-10 Ab is characterized by the PL band with a peak at 2.21 eV and a small PL shoulder in the range 2.3-3.0 eV (Figure 4, curve 1). The intensity of the PL band related to PBS (Figure 4, curve 2) is 15-fold smaller than

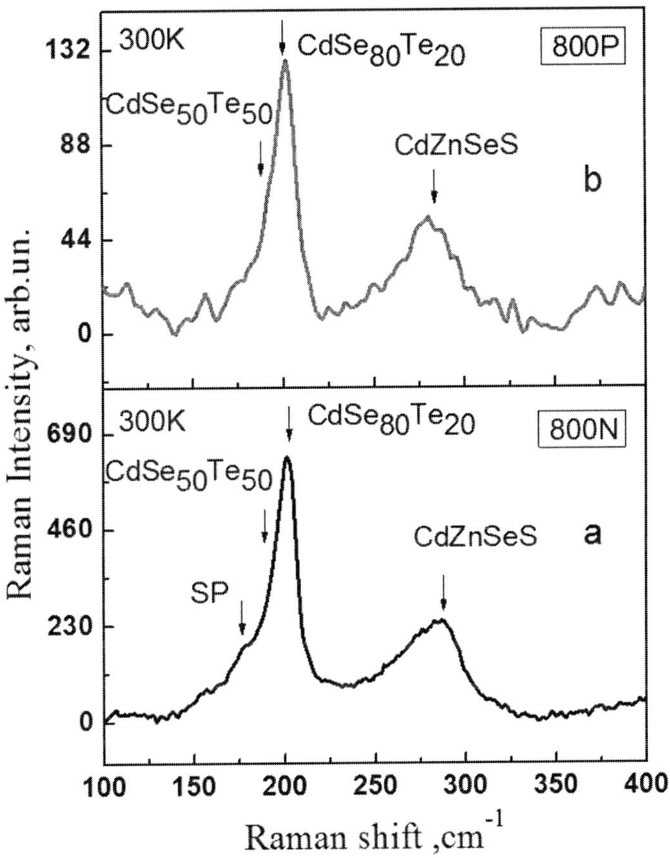

Figure 13. Raman scattering spectra of 800N (a) and 800P+ IL-10 (b) QDs.

the emission intensity of 2.21 eV PL band connected with anti-IL-10 Abs. Note that the 2.21 eV PL band intensity in anti-IL-10 Abs is 10-fold smaller than the emission intensity of excitons in CdSeTe/ZnS QDs at UV light excitation (Figure 11). Thus, the new PL bands peaked at 1.75 and 1.90 eV that appeared in the PL spectra of bioconjugated QDs do not related to the emission of anti-Il-10 Abs or PBS.

To study the impact of the excitation quanta energy on the shape of PL bands, the PL spectra of nonconjugated and bioconjugated QDs have been measured using the different excitation wavelengths of 532 nm, 488 nm, and 325 nm (Figure 12). The difference in PL intensities is related to the power of lasers used, but the shape of PL bands in the nonconjugated and bioconjugated QDs is similar. Note that the PL intensity of QD emission is higher at the UV excitation (325 nm).

The Raman scattering spectra of CdSeTe/ZnS QDs are investigated as well. The Raman spectrum of nonconjugated CdSeTe/ZnS QDs represents a complex Raman peak of QDs at 202.5 cm^{-1} and the Raman peak at 519.6 cm^{-1} connected with the first-order Raman scattering involving optic phonons in the Si substrate (Figure 13a). The Raman peak of QDs is a superposition of two Raman peaks at 191.2 and 202.5 cm^{-1} (Figure 13a), which correspond to the LO phonons in the $CdSe_{50}Te_{50}$ core and $CdSe_{80}Te_{20}$ core's cover layer [31, 50]. In addition, the Raman spectrum represents two peaks at 177.0 and 280 cm^{-1}. The peak at 177.0 cm^{-1} could be assigned to a surface phonon (SP) in QD cores [6]. The Raman peak at 280 cm^{-1} is probably due to the CdZnSeS alloy that appears at the CdSeTe/ZnS interface at the temperature of ZnS shell growth [6, 50-52].

The Raman scattering investigation shows (Figure 13a, b) that the Raman peaks related to the LO phonons in the double CdSeTe core and in the alloy at the interface did not vary at the QD bioconjugation. This fact testifies that the "blue" PL energy shift in bioconjugated CdSeTe/ZnS QDs (Figures 11 and 12) has been not connected with compressive strains or with the degradation of CdSeTe QD core owing to core/shell intermixing or core oxidation at the QD bioconjugation.

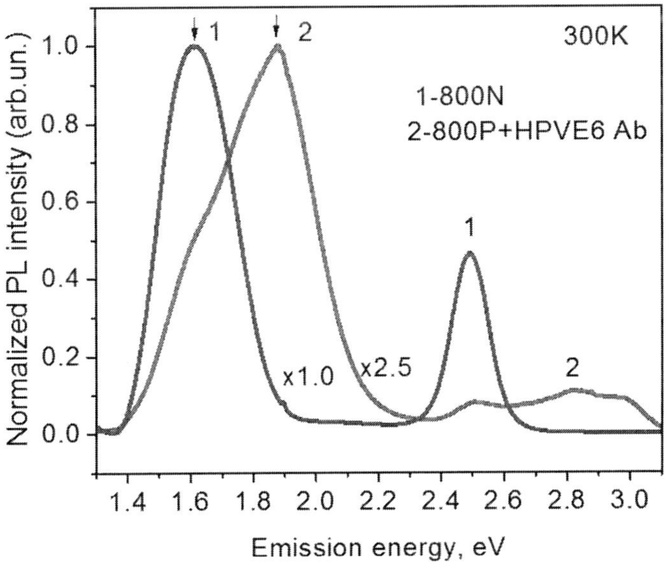

Figure 14. Normalized PL spectra of nonconjugated 800N (1) and bioconjugated 800P (2) QDs to HPV E6 Ab measured at 300 K. Numbers at the curves (×1.0 and ×2.5) present the multiplication coefficients used at the normalization of experimental PL spectra.

5.2. CdSeTe/ZnS QDs bioconjugated to the HPV and PRV antibodies

Normalized PL spectra of nonconjugated and bioconjugated CdSeTe/ZnS QDs have been presented in Figures 14, 15, and 16 for the different antibodies used: (i) mouse monoclonal anti-HPV E7 Ab, (ii) mouse monoclonal HPV E6 Ab, and (iii) pseudorabies virus (PRV) Ab.

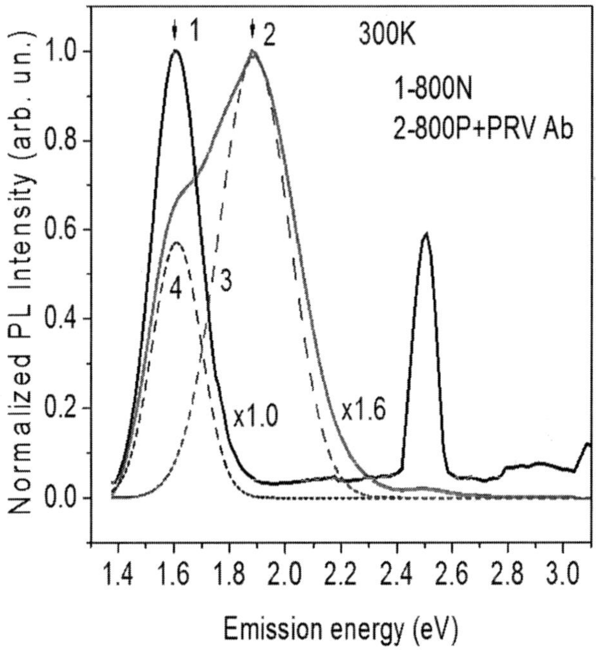

Figure 15. Normalized PL spectra of nonconjugated 800N (1) and bioconjugated 800P QDs (2) to PRV Ab measured at 300 K. Numbers (×1.0 and ×1.6) present the coefficients used at the normalization of PL spectra. Dashed curves (3 and 4) show the deconvolution of curve 2 on two PL bands.

QD emission spectra in nonconjugated state are characterized by the two Gaussian shape PL bands. The main PL band (I) with a maximum at 1.57-1.60 eV (Figures 14-16, curve 1) is connected with exciton emission in the CdSeTe core. The small intensity PL band (II) with a peak at 2.50 eV is attributed to exciton emission in the intermediate alloy CdZnSeS layer at the core/shell interface [50-52]. This intermediate layer was introduced at the process of QD growth with the aim to decreasing the mismatch between the CdSeTe and the ZnS crystal lattices. The PL intensity of 2.50 eV band varies owing to the variation, apparently, a volume of intermediate alloy layers in the different QD ensembles (Figures 14-16). The small intensity PL band at 2.50-2.90 eV (Figure 15) can be connected with the emission of PBS (see Figures 4 and 9) that exists in the nonconjugated QD ensembles as well.

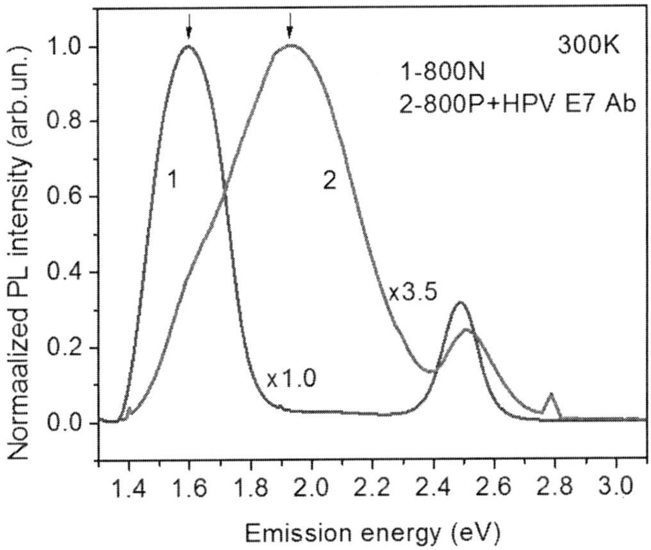

Figure 16. Normalized PL spectra of nonconjugated 800N (1) and bioconjugated 800P QDs (2) to HPV E7 Ab at 300 K. Numbers at the curves (×1.0 and ×3.5) correspond to multiplication coefficients used at the normalization of experimental PL spectra.

The main PL peak (I) shifts to higher energy in the PL spectra of bioconjugated QDs up to: 1.881 eV (Figure 14, curve 2), 1.887 eV (Figure 15, curve 2), or even 1.937 eV (Figure 16, curve 2). Simultaneously, the PL band shape becomes asymmetric with essential low-energy shoulders. The analysis has revealed that the main PL band in bioconjugated QDs is complex and can be represented by a superposition of two PL bands (Figure 15, curves 3 and 4): the low-energy band with a peak at 1.60 eV and the high-energy band centered at 1.887 eV in Figure 15. At the same time, the PL intensity of main PL band I decreases and the additional PL band appears in the spectral range 2.2-3.0 eV.

6. DISCUSSION OF THE CDSETE/ZNS QD BIOCONJUGATION

Let us discuss again the physical reasons of PL spectrum transformation at the QD bioconjugation to antibodies. As we mentioned above, these reasons can be related to (i) the emission of antibody molecules or PBS at UV excitation; (ii) the compressive strain applied to QDs at the bioconjugation [33]; (iii) the compound material degradation owing to core/shell intermixing [53] or oxidation [32] at the QD bioconjugation; (iv) the quantum-confined Stark effect [43, 44], stimulated by electric charges of antibodies; (v) the dominated emission of excitons localized at the excited QD states; (vi) the change of energy band profile at the application of electric field of charged Abs; and (vi) the quantum-confined effect owing to the shift of QD energy levels stimulated by the change of potential barrier at the QD surface [45] or by decreasing the effective QD size in bioconjugated QDs [46]. To make the decision concerning the most probable physical reasons of the PL transformation in CdSeTe/ZnS QDs, let us analyze all of these factors in details.

(i) The emission of HPV or PRV antibodies and PBS can be excited together with QD emission at the PL excitation by UV light (325 nm). The PL spectra of pure HPV and PRV antibodies, and PBS without QDs have been investigated (Figure 17).

The wide emission bands have been detected in the spectral range 2.0-3.2 eV in PL spectra of both the antibodies and PBS (Figure 17). The PL intensity of PBS emission is 50-fold smaller, and the PL intensities of antibody's emissions are 5- or 10-fold smaller in comparison with the PL intensity of the main PL bands I in bioconjugated CdSeTe/ZnS QDs. Thus, the emission of antibodies or PBS at the high-energy UV light excitation cans influent only on the shape of high-energy tails in PL spectra of bioconjugated QDs. This impact is seen clearly in Figures 14, 15, and 16 (curve 2), where the high-energy PL band appears in the spectral range 2.2-3.2 eV.

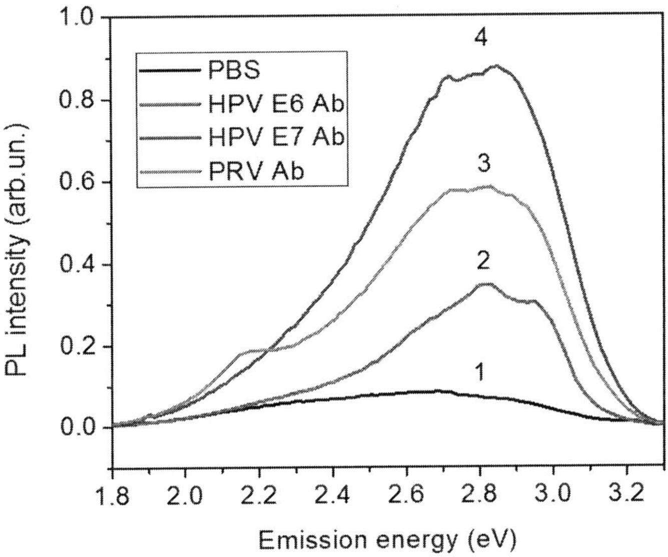

Figure 17. PL spectra of PBS (1), HPV E6 Ab (2), PRV Ab (3), and HPV E7 Ab (4) without QDs.

(ii-iii) If the compressive strain appears at QD bioconjugation, the Raman peaks has shift in Raman spectra in comparison with those for nonconjugated QDs. The Raman scattering spectrum of nonconjugated CdSeTe/ZnS QDs represents a complex Raman peak of QDs at 202.5 cm^{-1} and another Raman peak at 519.6 cm^{-1} related to the Si substrate (Figures 13a and 18a).

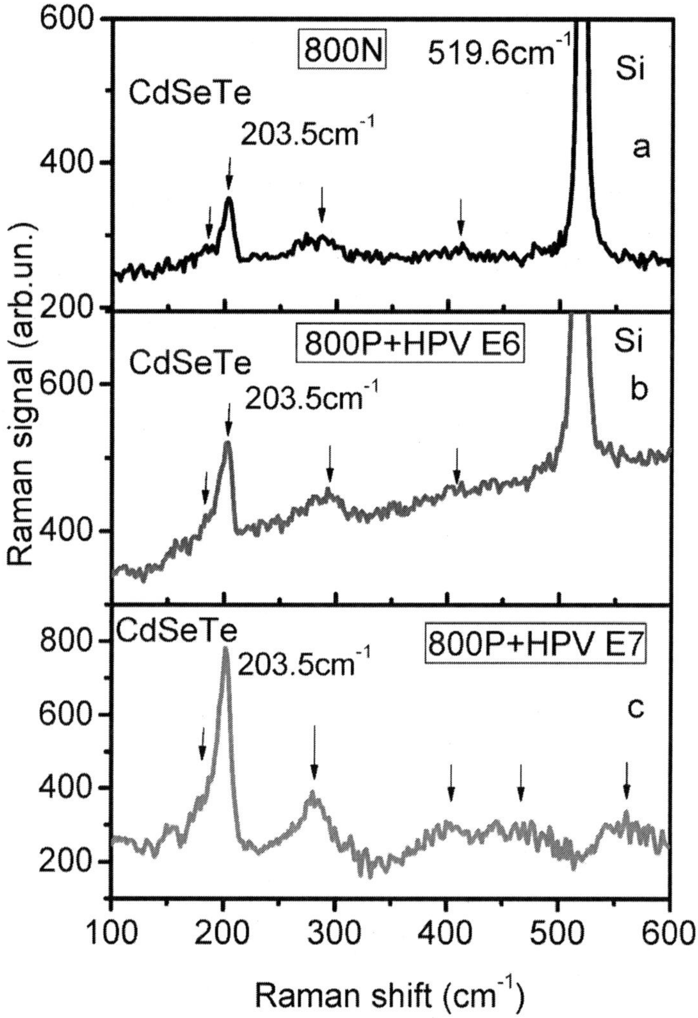

Figure 18. Raman scattering spectra of nonconjugated 800N (a) and bioconjugated 800P (b, c) QDs to HPV E6 (b) and HPV E7 (c).

The QD peak is a superposition of Raman peaks at 191.2 and 202.5 cm^{-1} (Figure 18a), which correspond to Raman scattering involving the LO phonons of the $CdSe_{50}Te_{50}$ core and $CdSe_{80}Te_{20}$ core's cover layer [31, 50]. The detailed estimations of material compositions in the double CdSeTe QD core on the base of Raman peak analysis were published earlier by Quintos Vazquez et al. [31]. The Raman spectrum demonstrates as well the peak at 280 cm^{-1}, which can be assigned to light scattering by LO phonons in the CdSeZnS intermediate alloy layer. The Raman peak at 406 cm^{-1} is related to two-phonon Raman scattering in the CdSeTe core [50]. The intensity of Raman signals in bioconjugated CdSeTe/ZnS QDs increases in comparison with those in nonconjugated QDs [54]. This effect was attributed early to the surface-enhanced Raman scattering (SERS) [22, 39, 54] due to the interaction of the electromagnetic field of excitation light with the electric charges (dipoles) of antibodies. The realization of the SERS effect at Raman scattering in bioconjugated QDs testifies that the antibodies have the electric charges.

The SERS effect is responsible for appearing the low-intensity Raman peaks (463 and 580 cm^{-1}) in Raman spectra of bioconjugated QDs (Figure 18c). These Raman signals are related to the Si substrate and were assigned early to scattering involving TO phonons (463 cm^{-1}) and the combination of TA and TO phonons at the X direction in the Si Brillouin zone [36]. The Raman scattering study shows (Figure 18b, c) that the positions of Raman peaks in Raman spectra of the double CdSeTe core and intermediate CdSeZnS layer have not been varied at the CdSeTe/ZnS QD bioconjugation. Permanent positions of Raman signals in QDs testify that the bioconjugation process has not connected with the compressive strains to QDs, with the compound degradation owing to core/shell intermixing or the QD core oxidation at the bioconjugation.

(iv) The quantum-confined "red" Stark energy shift was predicted theoretically for emission in the ensembles of CdS or CdSSe QDs [43, 44]. The quantum-confined Stark effect can be avoided from our discussion as well because the "blue" PL energy shift is detected in studied CdSeTe/ZnS QDs at the bioconjugation.

(v) To analyze the role of excitons bounded at the excited states in the CdSeTe core, the dependence of PL spectrum transformation versus excitation light intensities has been studied (Figure 19a-c). As it is clear from Figure 19a, the shape of PL bands in nonconjugated

CdSeTe/ZnS QDs (800N) has not changed at varying the excitation light intensities.

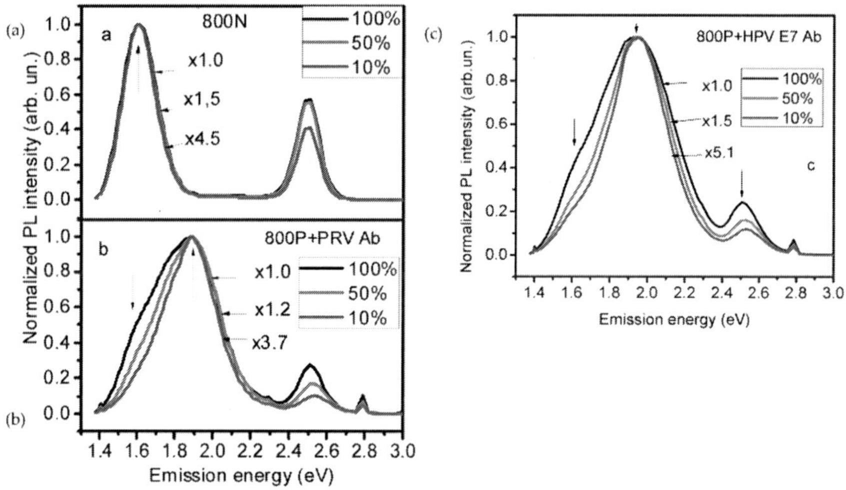

Figure 19. Normalized PL spectra of nonconjugated (a) and bioconjugated QDs to PRV Ab (b) and to HPV E7 Ab (c) measured at different excitation light powers at 300 K.

Meanwhile, the shape of the PL spectra of bioconjugated QDs varies versus excitation light power (Figure 19b, c). The PL intensity decreases mainly in the low-energy (1.60 eV) and high-energy (2.50 eV) ranges that have been seen clearly. The mentioned PL bands do not connect with the emission of excitons localized at the excited states in the CdSeTe core. Thus, the emission of excitons localized at the excited states in bioconjugated CdSeTe/ZnS QDs is not responsible for the transformation of PL spectra versus excitation light power.

(vi) Let us consider the energy band diagram of CdSeTe/ZnS QDs with the aim to analyze the varying energy band profile at the bioconjugation. The energy diagram of the double core CdSeTe/ZnS QDs with emission at 800 nm was calculated early by Quintos

Vazquez et al. [31]. The band gaps and electronic affinities for the $CdSe_{50}Te_{50}$, $CdSe_{80}Te_{20}$, and ZnS bulk crystals [31, 55] have been represented in Figure 20.

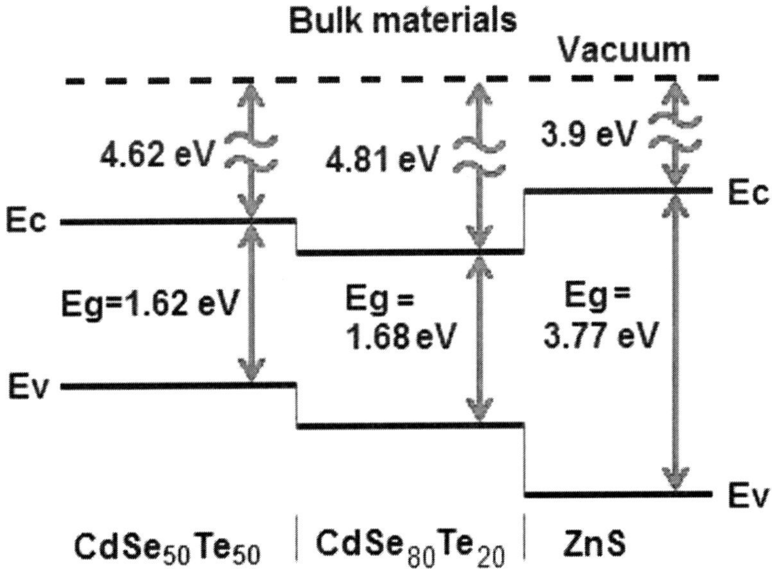

Figure 20. Energy band diagram for the CdSeTe alloys and ZnS bulk crystal.

To create the energy diagram of nonconjugated CdSeTe QDs (Figure 21), the variation of energy band gaps for the $CdSe_{50}Te_{50}$ core and $CdSe_{80}Te_{20}$ core cover layer with their sizes is estimated using the effective mass approximation model [31]. The big size of the double core/shell CdSeTe/ZnS QDs (emission, 800 nm) permits to apply the effective mass approximation approach [6].

The ground energy of electron-hole pairs and the energy of the first absorption transition are estimated as follows [56]:

$$E_{1S1S} = E_g + \frac{\hbar^2 \pi^2}{2\mu a^2} - 1.786\frac{e^2}{\varepsilon a} - 0.248R_y \qquad (5)$$

where E_g is the energy band gap of bulk material, α α is a QD radius, μ is the reduced mass, $\mu^{-1} = m^{*-1}_{e} + m^{*-1}_{h}$ μ−1=me*−1+mh*−1 , m_e* and m_h* are the effective masses of electrons and holes, and ε is the high-frequency permittivity. The term $e^2/\varepsilon\alpha$ e2εα in Eq. (5) describes the electron-hole Coulomb interaction, and the fourth term is a minor correction.

To analyze the energy band gap, the following parameters have been used for the bulk CdSe: E_g = 1.730 eV at 300 K, m_e* = 0.13 m_o, m_h* = 0.45 m_o, and the high-frequency permittivity ε = 8.2 [50, 51]. Using the parameters for the bulk CdTe: E_g = 1.50 eV at 300 K, m_e* = 0.11m$_o$, and m_h* = 0.35m$_o$ [50, 57], the values of energy band gaps and effective masses for the bulk $CdSe_{0.8}Te_{0.2}$ and $CdSe_{0.5}Te_{0.5}$ alloys have been obtained by the linear interpolation. To estimate the position of PL excitation peaks connected with the electron-hole ground states in QD, the value of Stokes shift is taken in account, which equals 50 meV [6]. Estimated energy band gaps for the $CdSe_{0.5}Te_{0.5}$ and $CdSe_{0.8}Te_{0.2}$ alloys have been used for the design of the energy diagram of the double core/shell QDs (Figure 21).

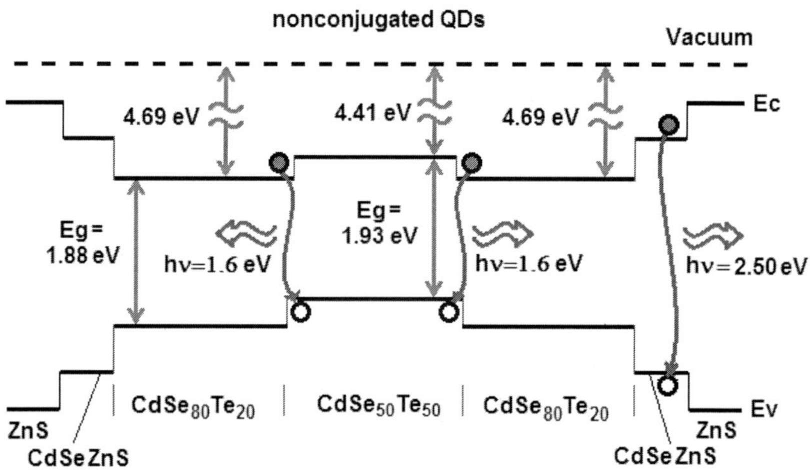

Figure 21. The energy band diagram of nonconjugated double core CdSeTe/ZnS QDs.

The energy band diagram of nonconjugated CdSeTe/ZnS QDs (Figure 21) explains the types of the optical transitions and emission energy of 1.60 eV detected in the PL spectra of nonconjugated QDs. As it is clear from Figure 21, owing to the type II quantum well ($CdSe_{80}Te_{20}/CdSe_{50}Te_{50}/CdSe_{80}Te_{20}$) in double core QDs, the wide band gap core and the core's covered quantum well layer with the energy band gaps of E_g = 1.93 and 1.88 eV, respectively, enable to emit IR emission with a PL peak at 1.60 eV in nonconjugated QDs (Figures 11, 14, 15, and 16). This 1.6-eV emission owes to the recombination of electrons, from the $CdSe_{80}Te_{20}$ conduction band, with holes localized in the $CdSe_{50}Te_{50}$ valence band (Figure 21).

To explain the PL band "blue" shift into the higher energy range (1.88-1.94 eV) in bioconjugated CdSeTe/ZnS QDs (Figures 11, 14, 15, and 16), let us to consider the transformation of band diagram at applying the external electric field of charged antibodies. In this case, the energy band profile varies as it is presented in Figure 22. Actually, in the bioconjugated QDs, the two types of optical transitions are possible: the recombination of excitons located at the ground states in the $CdSe_{50}Te_{50}$ core and/or the recombination of electrons, from the $CdSe_{80}Te_{20}$ conduction band, with holes localized in the $CdSe_{50}Te_{50}$ valence band (Figure 22).

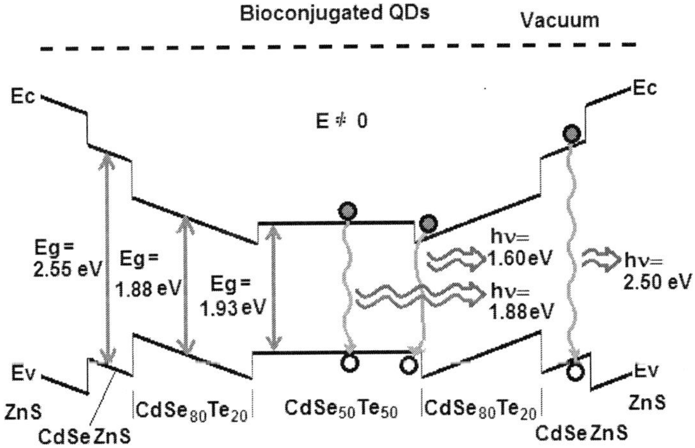

Figure 22. Energy band diagram of QDs bioconjugated to 800P+HPV E6 Ab.

The transformation of PL spectra versus excitation light intensities (Figure 19) has confirmed the proposed QD energy band diagrams. Actually, at low excitation light intensity, the recombination (1.88 eV) of excitons at the ground $CdSe_{50}Te_{50}$ core states dominates in the PL spectrum (Figure 22). At obtaining the last value (1.88 eV), the Stock shift of 50 meV is taken into account [6, 29, 30]. When excitation intensity increases, the indirect recombination transition (electrons from $CdSe_{80}Te_{20}$ with holes in $CdSe_{50}Te_{50}$) and exciton emission in the intermediate CdSeZnS layer can be realized. Hence, the energy band profile variation in the double core CdSeTe/ZnS QDs at the bioconjugation to charged antibodies permits to explain the PL spectrum peculiarities of bioconjugated QDs.

(vii) Finally, let us discuss the difference in PL band positions in CdSeTe/ZnS QDs bioconjugated to different antibodies (Figures 14-16, curve 2). The main PL peaks in bioconjugated QDs are as follows: 1.881 eV for 800P+HPV E6 Ab (Figure 14, curve 2), 1.887 eV for 800P+PRV Ab (Figure 15, curve 2), and 1.937 eV for 800P+HPV E7 Ab (Figure 16, curve 2). The difference in PL peaks can be related to the difference in electric potentials (potential barriers) at the surface of bioconjugated QDs. It is known that the position of energy levels in QDs, for the strong quantum confinement regime, depends on the value of potential barriers at the surface [45]. The comparison of PL intensities, detected for the pure antibodies (Figure 17), and the corresponding PL peaks in bioconjugated QDs (Figures 14-16, curve 2) have shown that the antibody HPV E7 with highest emission intensity (Figure 17, curve 4) stimulates the biggest PL band shift up to 1.937 eV (Figure 16, curve 2). The PL intensity, measured for the pure antibodies (Figure 17), depends on their concentrations in PBS. Thus, it is possible to conclude that the shift of PL spectra depends on a number of charged antibodies bioconjugated to CdSeTe/ZnS QDs [58]. Moreover, this PL shift increases by enlarging the antibody concentration in PBS. The potential barrier increasing at the QD surface has shifted the QD emission into the higher energy range owing to the shift of QD energy levels [58].

Thus, the PL spectrum transformation of CdSeTe/ZnS QDs, bioconjugated to studied antibodies, is related to the two effects: (i) the change of energy band profiles and (ii) the quantum-confined effect owing to the shift of QD energy levels stimulated by the change of electric potentials (potential barrier) at the surface of QDs bioconjugated to charged antibodies. The potential barrier increasing has shifted the QD energy levels and QD emission into the higher energy range.

7. CONCLUSION

Photoluminescence, its excitation power dependence, and Raman scattering spectra have been studied in CdSe/ZnS and CdSeTe/ZnS QDs for the nonconjugated states and after the QD conjugation to the anti-IL-10, human papilloma virus, and pseudorabies virus antibodies. The QD bioconjugation to charged antibodies stimulates the "blue" energy shift of PL bands related to exciton emission in the CdSe or CdSeTe cores.

The "blue" energy shift of PL spectrum in bioconjugated CdSe/ZnS QDs has been attributed to the electronic quantum-confined effects stimulated by decreasing the effective QD size at its bioconjugation to charged antibodies. It was shown that the attachment of a charge, deals with the antibody, to the exterior shell of CdSe/ZnS QDs leads to blocking away a fraction of core's volume.

The energy band diagrams of CdSeTe/ZnS QDs in the nonconjugated and bioconjugated states have been designed, which permit to explain the types of optical transitions in QDs and their transformations at the QD bioconjugation. It is shown that the change of energy band profile and the "blue" shift of QD energy levels, owing to the change of potential barrier at the QD surface, are the dominant reasons of PL spectrum transformation in the double core CdSeTe/ZnS QDs conjugated to charged antibodies. A better understanding of the QD bioconjugation to specific antibodies is expected to produce the major advances in biology and medicine and can be a powerful technique for early medical diagnostics.

8. ACKNOWLEDGEMENTS

The work was partially supported by CONACYT (project 130387) and by SIP-IPN (projects 20140648 and 20150601), Mexico. The author would like to thank Dr. Janna Douda for the QD bioconjugation to antibodies and Dr. Jose Luis Casas Espinola for the PL measurements.

REFERENCES

1. T. Jamieson, R. Bakhshi, D. Petrova, R. Pocock, M. Imani, A.M. Seifalian, Biomaterials 28, 4717 (2007).
2. D. Gerion, W.J. Parak, S.C. Williams, D. Zanchet, C.M. Micheel, A.P. Alivisatos, J Am. Chem. Soc. 124, 7070 (2002).
3. X.H. Gao, Y.Y. Cui, R.M. Levenson, L.WK. Chung, S.M. Nie, Nat. Biotechnol. 22, 969 (2004).
4. Z. Liu, W.B. Cai, L.N. He, N. Nakayama, K. Chen, X.M. Sun, X.Y. Chen, H.J. Dai, Nat. Nanotech. 2, 47 (2007).
5. R. Weissleder, K. Kelly, E.Y. Sun, T. Shtatland, L. Josephson, Nat. Biotechnol. 23, 1418 (2005).
6. T. Torchynska, Yu. Vorobiev. Semiconductor II-VI Quantum Dots with Interface States and Their Biomedical Applications, Advanced Biomedical Engineering, Gaetano D. Gargiulo and Alistair McEwan (Eds.), InTech Publisher, Croatia, pp. 143-182, 2011.
7. L.J. Yang, Y.B. Li, Analyst 131, 394 (2006).
8. I. Brigger, C. Dubernet, P. Couvreur, Adv. Drug Deliv. Rev. 54, 631 (2002).
9. Y. Ebenstein, T. Mokari, U. Banin, J. Phys. Chem. B 108, 93 (2004).
10. F. Wang, W.B. Tan, Y. Zhang, X. Fan, M. Wang, Nanotechnology 17, R1-R13 (2006).

11. M. Kuno, D.P. Fromm, H.F. Hamann, A. Gallagher, D.J. Nesbitt, J. Chem. Phys.115, 1028 (2001).

12. A.R. Clapp, I.L. Medintz, J.M., Mauro, Br.R. Fisher, M.G. Bawendi, H. Mattoussi, J. Am. Chem. Soc. 126, 301 (2004).

13. N. Tessler, V. Medvedev, M. Kazes, S.H. Kan, U. Banin, Science 295, 1506 (2002).

14. Th.E. Rasmussen, L. Jauffred, J. Brewer, S. Vogel, E.R. Torbensen, B.Ch. Lagerholm, L. Oddershede, E.C. Arnspang, J. Mod. Phys. 4, 27 (2013).

15. K. Koenig, J. Microsc. 200, 83 (2000).

16. H.M.E. Azzazy, M.M.H. Mansour, S.C. Kazmierczak, Clin. Biochem. 40, 917 (2007).

17. C.H. Lin, L.W Chang, H. Chang, M.H. Yang, Ch.Sh. Yang, W.H. Lai, W.H. Chang, P. Lin, Nanotechnology 20, 215101 (2009).

18. B.A. Fowler, Toxicol. Appl. Pharmacol. 238, 294 (2009).

19. W.J. Parak, D. Gerion, D. Zanchet, A.S. Woerz, T. Pellegrino, Ch. Micheel, Sh. C. Williams, M. Seitz, R.E. Bruehl, Z. Bryant, C. Bustamante, C.R. Bertozzi, A. P. Alivisatos, Chem. Mater. 14, 2113 (2002).

20. A. Wolcott, D. Gerion, M. Visconte, J. Sun, Ad. Schwartzberg, Sh. Chen, J. Z. Zhang, J. Phys. Chem. B 110, 5779 (2006).

21. D. Gerion, F. Pinaud, Sh.C. Williams, W.J. Parak, D. Zanchet, Sh. Weiss, A. P. Alivisatos, J. Phys. Chem. B 105, 8861 (2001).

22. T.V. Torchynska, Nanotechnology 20, 095401 (2009).

23. X. Ji, J. Zheng, J. Xu, V.K. Rastogi, T.Ch. Cheng, J.J. DeFrank, R.M. Leblanc, J. Phys. Chem. B 109 (2005) 3793.

24. L.G. Vega Macotela, J. Douda, T.V. Torchynska, R. Peña Sierra, L. Shcherbyna, Phys. Status Solidi (c) 7, 724 (2010).

25. W. Guo, J. Jack Li, J.A. Wang, X. Peng, Chem. Mater. 15, 3125 (2003).

26. T.V. Torchynska, J. Douda, P.A. Calva, S.S. Ostapenko, R. Peña Sierra, J. Vac. Sci. Technol. 27(2), 836 (2009).

27. T.Y. Ohulchanskyy, A.M. Pliss, P.N. Prasad, Biophotonics: harnessing light for biology and medicine, B. Di Bartolo and J. Collins (Eds.), Biophotonics: Spectroscopy, Imaging, Sensing, and Manipulation, Springer Science+Business Media B.V., 2011, p. 3.

28. T.V. Torchynska, J. Douda, S.S. Ostapenko, S. Jimenez-Sandoval, C. Phelan, A. Zajac, T. Zhukov, T. Sellers, J. Non-Crystal. Solids, 354, 2885 (2008).

29. http://www.lifetechnologies.com/ua/en/home/brands/invitrogen.html

30. [30]http://www.lifetechnologies.com/ua/en/home/life-science/cell-analysis/qdots-microspheres-nanospheres.html?ICID=ta-lm-quantumdots-QuantumDot&Microspheres

31. A.L. Quintos Vazquez, T.V. Torchynska, J.L. Casas Espinola, J.A. Jaramillo Gómez, J. Douda, J. Luminesc. 143, 38 (2013).

32. A.Y. Nazzal, X. Wang, L. Qu, W. Yu, Yu. Wang, X. Peng, M. Xiao, J. Phys. Chem. B 108, 5507 (2004).

33. R.W. Meulenberg, T. Jennings, G.F. Strouse, Phys. Rev. B 70, 235311 (2004).

34. M. Dybiec, G. Chomokur, S. Ostapenko, A. Wolcott, J.Z. Zhang, A. Zajac, C. Phelan, T. Sellers, G. Gerion, Appl. Phys. Lett. 90, 263112 (2007).

35. X. Gao, W.C.W. Chan, Sh. Nie, J. Biomed. Opt. 7(4), 532 (2002).

36. P.A. Temple, C.E. Hathaway, Phys. Rev. B 7, 3685 (1973).

37. F.A. Johnson, R. Loudon, Proc. R. Soc. A 281, 274 (1964).

38. N.E. Korsunskaya, I.V. Markevich, T.V. Torchinskaya, M.K. Sheinkman, Phys. Status Solidi (a) 60, 565 (1980).

39. T.V. Torchynska, A. Diaz Cano, M. Dybic, S. Ostapenko, M. Morales Rodrigez, S. Jimenes Sandoval, Y. Vorobiev, C. Phelan, A. Zajac, T. Zhukov, T. Sellers, Phys. Status Solidi (c) 4, 241 (2007).

40. T.V. Torchynska, J. Douda, R. Peña Sierra, Phys. Status Solidi (c) 6, S143 (2009).

41. A. Diaz Cano, S. Jiménez Sandoval, Y. Vorobiev, F. Rodriguez Melgarejo, T. V. Torchynska, Nanotechnology 21, 134016 (2010).

42. V. Torchynska, J.L. Casas Espinola, A. Díaz Cano, J. Douda, K. Gazarian, Physica E, 51, 60 (2013).

43. S.A. Empedocles, M.G. Bawendi, Science 278, 2114 (1997).

44. D.J. Norris, A. Sacra, C.B. Murray, M.G. Bawendy, Phys. Rev. Lett. 72, 2612 (1994).

45. D.B. Tran Thoai, Y.Z. Hu, S.W. Koch, Phys. Rev. B 41, 6079 (1990).

46. T.V. Torchynska, Y.V. Vorobiev, V.P. Makhniy, P.P. Horley, Physica B, 453 68 (2014).

47. P.P. Horley, P. Ribeiro, V.R. Vieira, J. González-Hernández, Yu.V. Vorobiev, L.G. Trápaga-Martínez, Physica E 44, 1602 (2012).

48. D.J. Norris, M.G. Bawendi, Phys. Rev. B 53, 16338 (1996).

49. C. Kittel, Introduction to Solid State Physics, 6th ed., John Wiley, NewYork,1986. 1986.

50. Physics of II-VI compounds, A.N. Georgobiany and M.K. Sheinkman (Eds.), Publisher "Nauka," Moscow, Russia, 1986, 300 p.

51. Physics and chemistry of II-VI compounds, M. Aven, J.S. Prener (Eds.), North-Holland, Amsterdam, 1967, p. 625.

52. T.V. Torchynska, J. Luminesc. 137, 157 (2013).

53. X. Xia, Z. Liu, G. Du, Y. Li, M. Ma, J. Phys. Chem. C 114, 13414 (2010).

54. T.V. Torchynska, J.L. Casas Espinola, J.A. Jaramillo Gómez, J. Douda, K. Gazarian Physica E, 51, 55 (2013).

55. A.G. Milnes, D.L. Feucht, Heterojunctions and Metal-Semiconductor junctions, Academic Press, New York, 1972.

56. Y. Kayanuma, Phys. Rev. B. 38, 9797 (1988).

57. N.E. Korsunskaya, I.V. Markevich, T.V. Torchinskaya, M.K. Sheinkman, J. Phys. Chem. Solids 43, 475 (1982).

58. T.V. Torchynska, Physica E 68, 87 (2015).

9

CHAPTER

GROWTH OF INAS QUANTUM DOTS ON GERMANIUM SUBSTRATE USING METAL ORGANIC CHEMICAL VAPOR DEPOSITION TECHNIQUE

Tanuj Dhawan Æ Renu Tyagi Æ Rajesh Kumar Bag Æ Mahavir Singh Æ Premila Mohan Æ T. Haldar Æ R. Murlidharan Æ R. P. Tandon

(1)Department of Physics & Astrophysics, University of Delhi

(2)Solid State Physics Laboratory

ABSTRACT

Self-assembled InAs quantum dots (QDs) were grown on germanium substrates by metal organic chemical vapor deposition technique. Effects of growth temperature and InAs coverage on the size, density, and height of quantum dots were investigated. Growth temperature was varied from 400 to 450 °C and InAs coverage was varied between 1.40 and 2.35 monolayers (MLs). The surface morphology and structural characteristics of the quantum dots analyzed by atomic force microscope revealed that the density of the InAs quantum dots first increased and then decreased with the amount of InAs coverage; whereas density decreased with increase in growth temperature. It was observed that the size and height of

InAs quantum dots increased with increase in both temperature and InAs coverage. The density of QDs was effectively controlled by growth temperature and InAs coverage on GaAs buffer layer.

Keywords: Quantum dots Ge substrate InAs Self-assembled

INTRODUCTION

Self-assembled InAs quantum dots (QDs) grown on GaAs substrate have generated a momentous interest in the last few years due to their potential applications in QD lasers [1, 2] and infrared photodetectors [3, 4]. QD lasers have demonstrated superior characteristics such as ultra-low-threshold current densities [5, 6], ultra-high-temperature stability [7], very high differential efficiency [8], small α-factor, and correspondingly reduced filamentation [9]. GaAs has been the most common substrate used for the growth of InAs quantum dots although germanium (Ge) is equally interesting and promising alternative substrate material because of its almost identical lattice constant and thermal expansion coefficient to those of GaAs. Ge is cost effective and is available in large wafer sizes. Ge has higher mechanical strength than GaAs and therefore thinner Ge substrate can be used, resulting in light weight end products [10]. For GaAs substrates etch pit density (EPD) value is generally above $100/cm^2$ but Ge substrates are of a perfect crystalline quality (0 EPD) [11]. This makes Ge a suitable replacement of GaAs in various above-mentioned applications.

InAs QDs grown on GaAs substrates by both molecular beam epitaxy (MBE) and metal organic chemical vapor deposition (MOCVD) techniques have been reported by several groups [12, 13], and effects of growth rate, V/III ratio, growth temperature, and material coverage on InAs QDs grown on GaAs substrate have been well studied [14]. Ge substrates have also been used for the fabrication of InGaAs/AlGaAs lasers [11] and (Al)GaInP multi-quantum well LEDs [15]. Surprisingly, there is only one report by Knuuttila et al. [16] describing the growth of non-uniformly distributed InAs islands directly on Ge substrate without any buffer layer. In the growth of InAs QDs directly on Ge substrate by Stranski–Krastanow (SK) growth mode, the formation of anti phase domains (APDs) is inevitable because the dot formation takes place after a thin wetting layer is grown. This thin wetting layer itself will be infested with defects. This may adversely influence the growth of good quality dots subsequently. This could be one of the possible reasons which did not allow other groups to have a successful growth of InAs quantum dots directly on Ge substrates. Moreover, APDs may be reduced to a maximum extend if a buffer layer is grown on Ge substrate. Buffer layer assists in reducing the surface defects of the substrate and helps in confining InAs QDs. However, growth of uniformly distributed InAs QDs on Ge substrate using GaAs buffer layer is scarcely reported. Motivated by the above ideas, we report the effect of growth temperature and InAs coverage on the formation of InAs QDs on Ge substrate having a GaAs buffer layer.

EXPERIMENTAL DETAILS

Growth of GaAs and InAs layers was carried out using Thomas Swan horizontal MOCVD reactor. The source materials used were TMGa, TMIn, and AsH_3, and Ge (001) 6° off toward (110) substrates were used for all the growth runs. GaAs buffer layer of thickness 360 nm was grown at 660 °C and at a reactor pressure of 660 Torr. InAs quantum dots were grown for different InAs coverage and temperature. Growth temperature was varied from 400 to 450 °C and InAs coverage was varied between 1.40 and 2.35 monolayers (MLs). V/III ratio was kept at 90. Prior to growth, Ge substrates were degreased with organic solvent, then etched by an etchant $HF:H_2O_2:H_2O$ (1:1:5) for 2 mins, and finally etched in dilute HF to remove surface oxide. Prior to loading into the reactor, the substrates were dried by blowing dry N_2.. Veeco Nanoscope III atomic force microscope (AFM) was used to study the topography of the grown QDs. Dot size, density, and height were also determined as a function of temperature and growth time. Crystalline quality of GaAs buffer layer grown on Ge substrate was studied using double crystal X-ray diffraction (X'Pert MRD, Philips). Optical characterizations of the samples were performed by photoluminescence (PL) measurement setup (Dongwoo) at a temperature of 77 K, using a diode laser of wavelength 532 nm. The PL signal was detected by a Si detector.

RESULTS AND DISCUSSIONS

When polar material (GaAs, InAs) is epitaxially grown on non-polar material (Ge), it often leads to structural defects known as APDs which creates deep levels inside the forbidden band and acts as strong scattering centers [17, 18]. They can also result in significant surface roughness which hinders large area uniformity. The most common methods for avoiding APDs from GaAs/Ge interfaces is by deliberately using misoriented substrates [19–22] and growth using arsenic pre-layer [23]. Thickness of GaAs buffer layer is also crucial, as antiphase domains self-annihilate in a region of about 50 nm from the interface, leaving the GaAs final surface nearly free of APDs [21].

In the present study we have followed a two-step growth process to grow GaAs buffer layer, the details on growth are reported elsewhere [10]. We have used 6° off Ge substrate, grown with a GaAs buffer layer of thickness 360 nm using arsenic pre-layer. In order to ascertain crystalline quality of GaAs buffer layer grown on Ge substrate, X-ray rocking curve was studied using double crystal X-ray diffraction (Fig. 1). The CuK_{α} (λ = 1.54 Å) radiation was used as a source of radiation and (004) reflection was studied. Full width at half maximum (FWHM) of GaAs was found to be 36 arc sec, which is less than or comparable to the earlier reports [10, 24]. The narrowness of the FWHM of the GaAs epitaxial film peak indicates that the crystalline quality of GaAs epitaxial film is reasonably good.

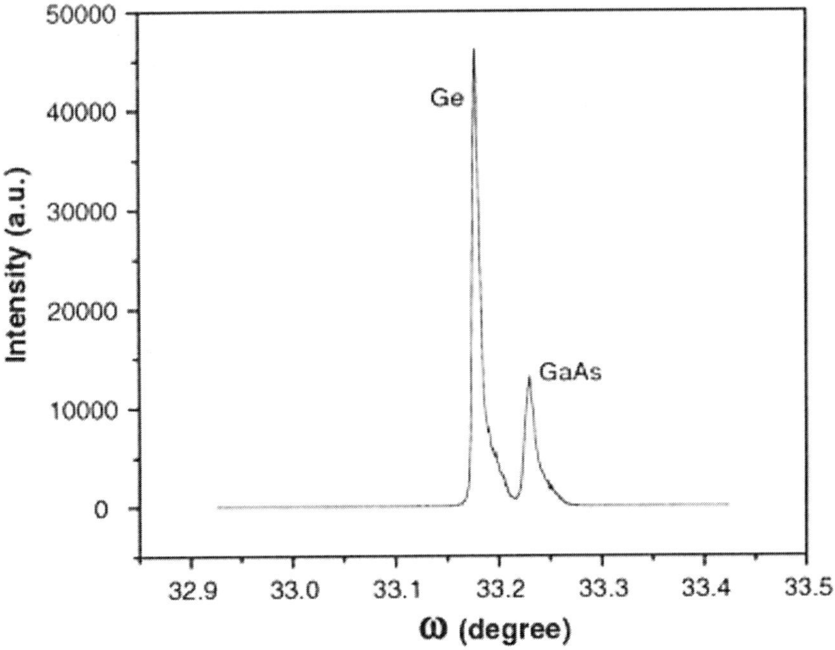

Figure 1. DCXRD rocking curve from the (400) Bragg lines of GaAs epitaxial films grown at 660 °C on Ge substrate

In first set of experiments, growth temperature for InAs QDs was fixed at 420 °C and InAs coverage was varied between 1.40 and 2.35 MLs. Figure 2 shows the AFM images of QDs at different InAs coverage of (a) 1.4 MLs, (b) 1.80 MLs, (c) 1.85 MLs, (d) 1.95 MLs, and (e) 2.35 MLs. In SK growth mode, few monolayers of two-dimensional (2-D) growth occur prior to the QDs formation. The critical layer thickness at which this 2-D to 3-D transition occurs depends on the lattice mismatch between the layer being deposited and the substrate. For InAs layer deposited on GaAs substrate, the critical layer thickness is around 1.6 MLs. Figure 2a shows the surface morphology of sample with coverage of 1.4 MLs. At the lowest material coverage of 1.4 MLs, there is no signature of quantum dots and only steps on the Ge substrate are visible. This is because critical thickness has not been achieved and 2-D wetting layer is present. As we increase deposition thickness to 1.8 MLs which is above the critical thickness, QDs formation was observed with very low density of dots.

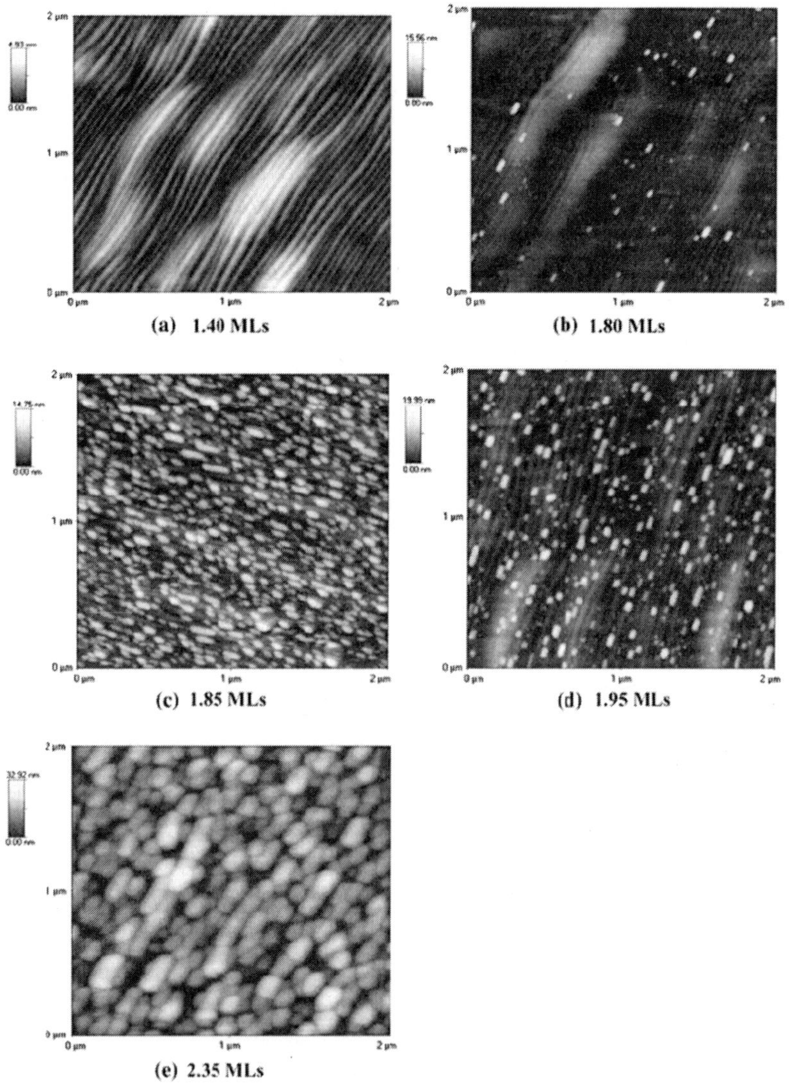

Figure 2. AFM images of InAs QDs grown at 420 °C on Ge substrate with InAs coverage of **A** 1.4 MLs, **B** 1.80 MLs, **C** 1.85 MLs, **D** 1.95 MLs, and **E** 2.35 MLs. The scan size for all the images is 2 μm × 2 μm

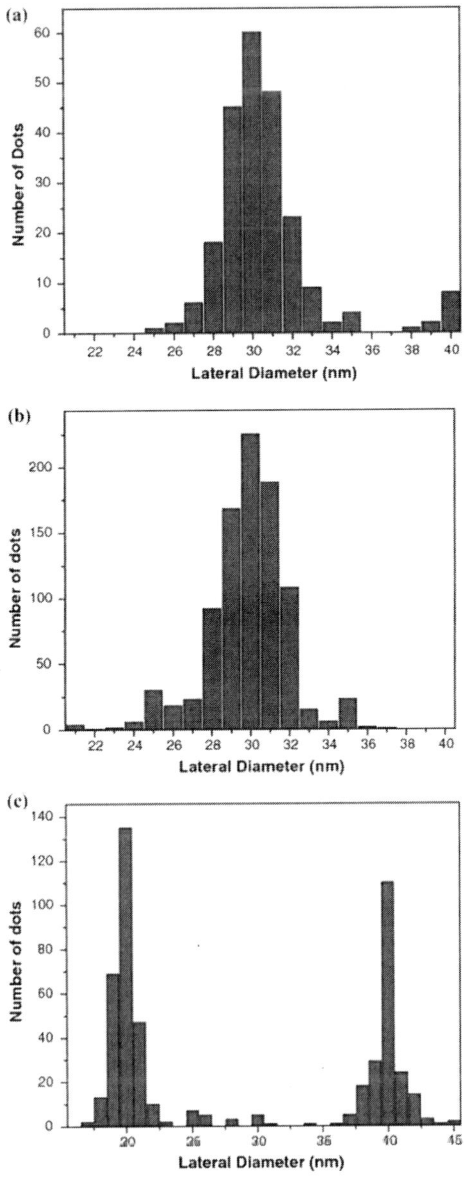

Figure 3. Histograms of the InAs quantum dots grown at 420 °C and InAs coverage of **A** 1.80 MLs, **B** 1.85 MLs, **C** 1.95 MLs showing variation of number of dots with lateral diameter. These histograms were created from 2 µm × 2 µm AFM images

The samples with InAs coverage of 1.80 and 1.85 MLs show nearly unimodal size distribution of the dots having a dot diameter and average height of about 30 nm ± 5% and 3.0 nm ± 5%, 30 nm ± 5% and 3.9 nm ± 5%, respectively. However, the sample with InAs coverage of 1.95 MLs shows bi-modal size distribution with smaller dots having 3.3 nm ± 5% average height and 20 nm ± 5% diameter while larger dots having 5.2 nm ± 5% average height and 40 nm ± 5% diameter. Fig. 3a, 3b confirms that the samples with InAs coverage of 1.80 and 1.85 MLs have mono-modal size distribution whereas sample with InAs coverage of 1.95 MLs (Fig 3c) have bi-modal size distribution. In the sample shown in Fig. 2e, with InAs coverage of 2.35 MLs, the increased InAs coverage has led to coalescence for the formation of bigger islands.

Density of dots is determined from AFM images of Fig. 2. A linear growth in height of QDs was observed as a function of InAs coverage as depicted in Fig. 4a. In the initial stage of growth only very few small dots formed with large separation as can be seen from Fig. 2b resulting in a low density. As InAs coverage increases, nucleation proceeds, density of dots increases, but dots are still well separated, as in Fig. 2c. The further growth in dot size leads to decrease in density due to coalescence of QDs as shown in Fig. 4b [also, refer Fig. 2d]. The densities of QDs for InAs coverage of 1.80, 1.85, 1.95, and 2.35 MLs are 5.5×10^{9}, 2.28×10^{10}, 1.27×10^{10}, and 7.43×10^{9} cm^{-2}, respectively. At coverage of 1.85 MLs, the size and density of the dots seem to be optimum for various possible device applications. Similar trend has been reported for InAs QDs grown on GaAs substrate [25].

The emission characteristics of the InAs QDs were determined by measuring their PL spectra at 77 K. Representative PL spectra of the sample having maximum density of dots are shown in Fig. 5. The observed PL spectra were quite broad and very much similar to that shown by Knuuttila et al. [16], although the peak energies are slightly different. Therefore, we further analyze the PL spectra by deconvolution as fitted to three distinct emission peaks (1.37, 1.33, and 1.29 eV) shown in Fig. 5. The most intense peak has the energy equal to the average of other two peak energies. The deconvolution PL spectra reflect an energy shift of the main peak by ±0.04 eV. This shift can easily be attributed to a marginal variation in dot sizes.

However, it should be noted that the blue shift of QD band gap (and hence the emission energy) due to the quantum confinement is inversely proportional to the square of dot size and the effective mass of the charge carriers of the material. Since effective mass of electron in InAs is quite small, a large blue shift is expected even for small change in dot size. The deviation of peak energy by ±0.04 eV corresponds to a deviation of dot size by ±0.1 nm for InAs in this energy range. The PL peak energy (1.33 eV) observed in the present study is greater than the earlier reported value of 1.15 eV which indicates the smaller size of QDs in our case [16]. PL peak energies of other samples were found to be consistent with their corresponding dot sizes.

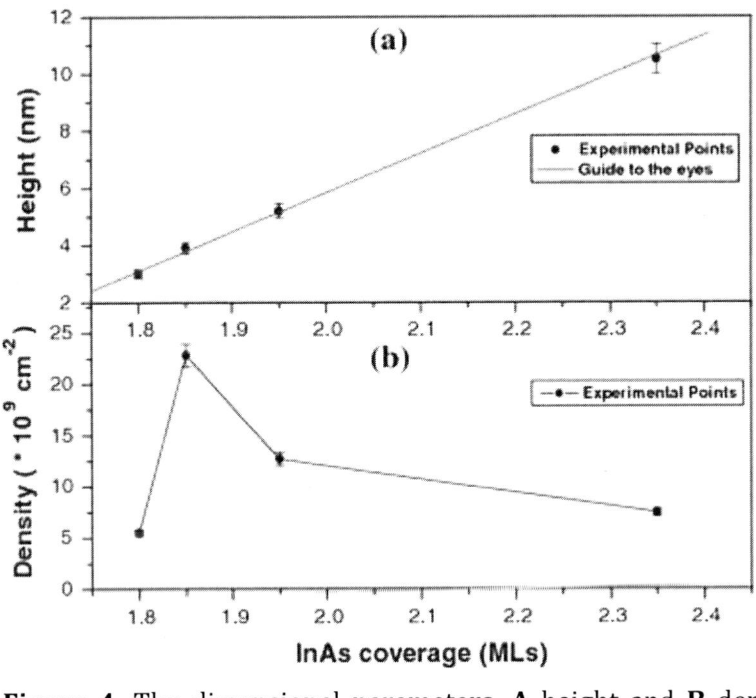

Figure 4. The dimensional parameters, **A** height and **B** density of quantum dots as a function of InAs coverage at 420 °C

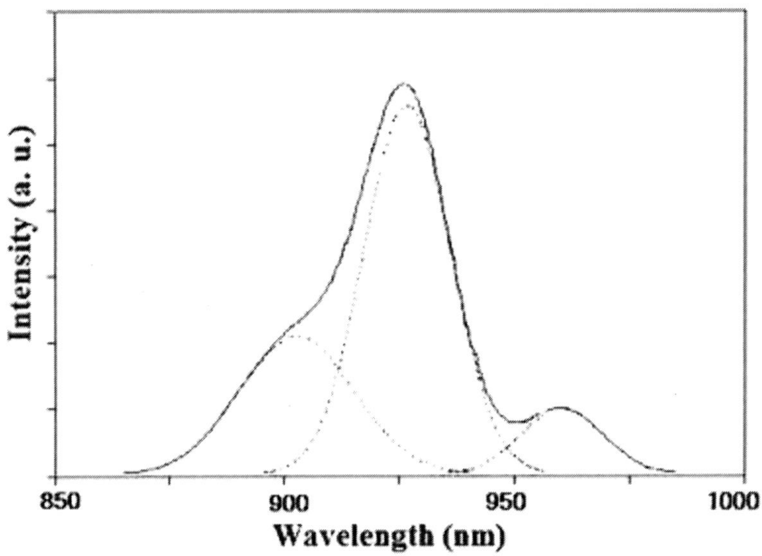

Figure 5. PL spectra of QDs grown at 420 °C and under InAs coverage of 1.85 MLs measured at 77 K

Figure 6 shows the PL spectra of QDs grown at 420 °C with InAs coverage of 1.4 MLs and 1.8 MLs measured at 77 K. PL spectra of sample with InAs coverage of 1.4 MLs do not show any sign of QD formation. Only GaAs peak is visible in the spectrum. PL spectra with InAs coverage of 1.8 MLs show the QD peaks at 1.31, 1.34, and 1.36 eV. In this energy region, all three QD peaks represent almost same size. Comparing PL spectra of samples with InAs coverage of 1.80 MLs and 1.85 MLs, we find the peak positions in PL spectra of both the samples are almost same. This shows QD size is same in both the samples. This is also confirmed by the AFM studies on these samples. Both the samples differ in the density of QDs. Sample with more InAs coverage has greater density of QDs.

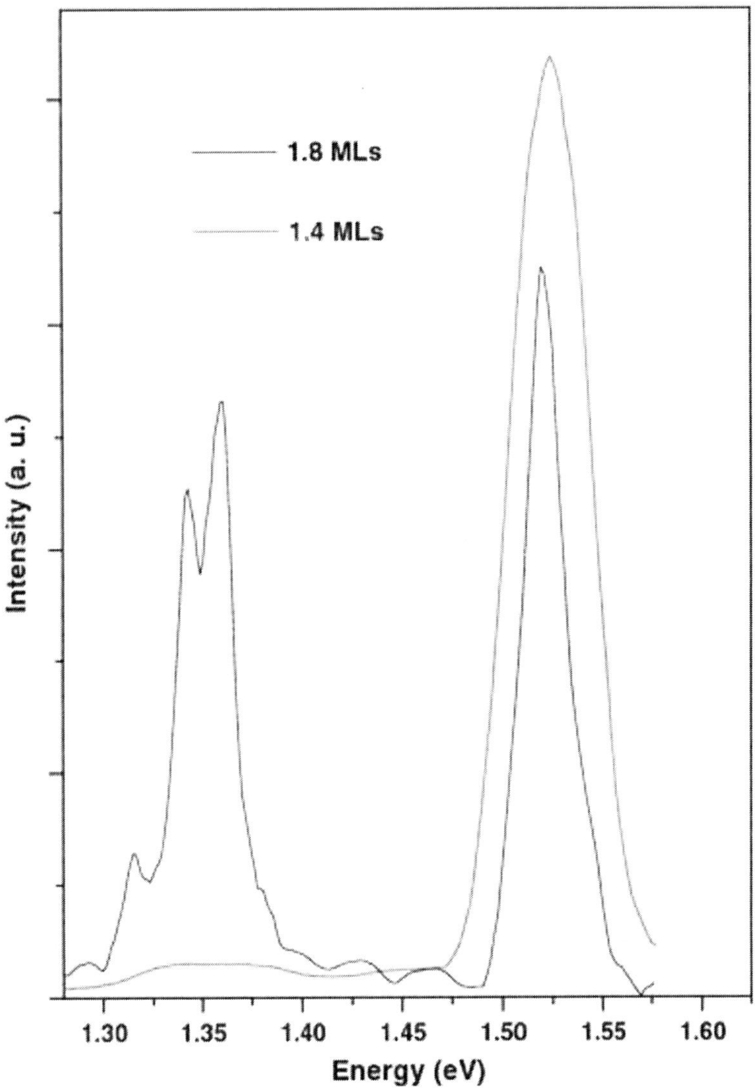

Figure 6. PL spectra of QDs grown at 420 °C with InAs coverage of 1.4 MLs and 1.8 MLs mcasured at 77 K

Temperature is one of the key parameters in the formation of QDs. It affects the adatom mobility and hence the density and size of the QDs. InAs coverage was fixed at 1.95 MLs and effect of growth temperature was investigated in the temperature range of 400 to 450 °C. For sample grown at 400 °C (Fig. 7a), shallow islands with base diameter of 40 nm ± 5% were formed. As we increased the temperature, the dots were found to change their shape significantly. At a temperature of 420 °C, spherical QDs were formed. On careful examination we found that dot height has increased with the increase in temperature. Deposited material has redistributed itself to form spherical dots (Fig. 7b). On further increase in temperature (Fig. 7c) large islands are formed due to increased surface adatom mobility.

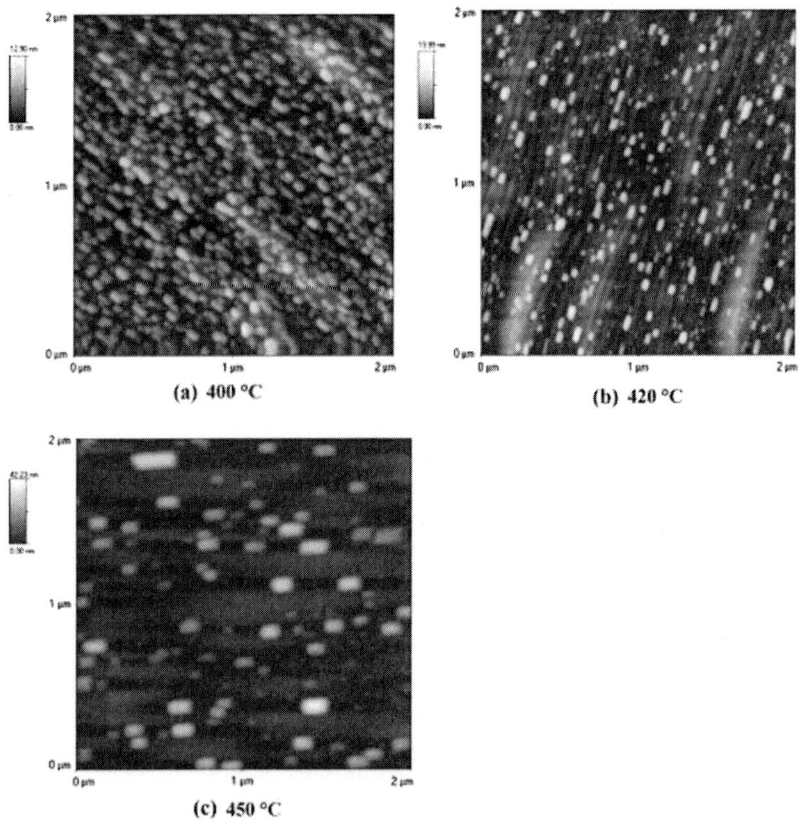

Figure 7. AFM images of InAs QDs grown at **A** 400 °C, **B** 420 °C, **C** 450 °C, with InAs coverage of 1.95 MLs. The size of each scan is 2 µm × 2 µm

Density of QDs decreased with increase in temperature due to coalescence of dots. Dots have coalesced due to increased adatom mobility with increase in temperature. The density of dot at 400, 420, and 450 °C was found to be 2.18 × 10^{10}, 1.27 × 10^{10}, and 3.75 × 10^{9} cm^{-2}, respectively. Variation of dot height and density with temperature is shown in Fig. 8. Dot height increased with the increase in temperature. The dots of average height 3.1 nm ± 5%, 3.3 nm ± 5%, and 4.9 nm ± 5% were grown at temperatures 400, 420, and 450° C, respectively. Similar variation of height and density of InAs QDs on GaAs substrate with temperature has been reported by other authors [26].

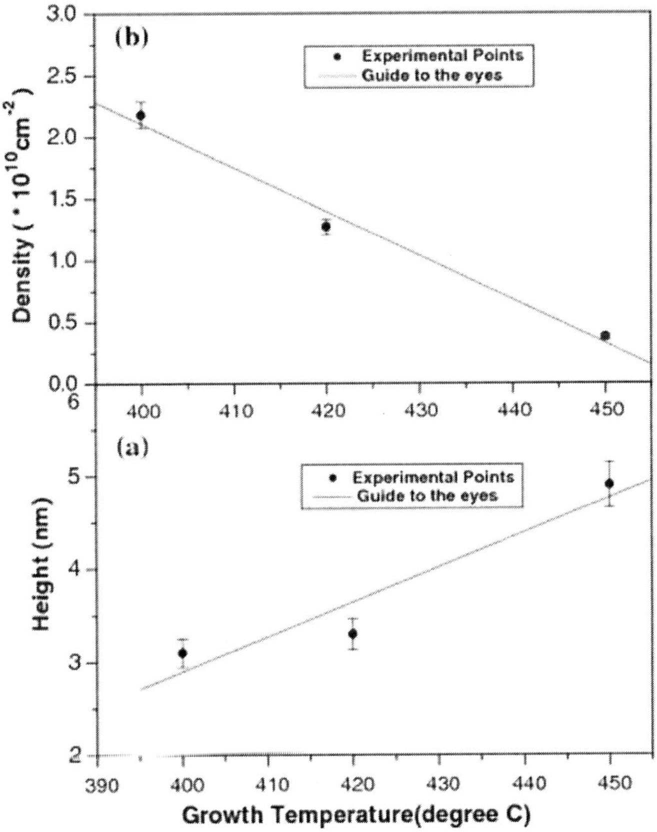

Figure 8. The dimensional parameters, **A** height and **B** density of quantum dots as a function of growth temperature for a InAs coverage of 1.95 MLs

CONCLUSIONS

Self-assembled InAs QDs were grown on Ge substrates using a GaAs buffer layer. Effects of InAs coverage and growth temperature on size, height, and density of InAs QDs on Ge substrate were investigated. Height and size of the QDs increased by increasing both InAs coverage and growth temperature. Density of QDs decreased with increase in growth temperature whereas in case of InAs coverage, density of dots first increased and then decreased with increase in InAs coverage. Density of QDs was effectively controlled by varying its growth temperature and amount of material deposited. The growth of good quality InAs quantum dots on Ge substrates for possible device applications is demonstrated. Hence, Ge can be used as an alternative substrate to GaAs for the successful growth of InAs QDs.

ACKNOWLEDGMENTS

Authors are thankful to Mr. Suresh Jangir, SSPL, New Delhi, for PL measurements and Dr. A. G. Vedeshwar, Department of Physics & Astrophysics, University of Delhi, for valuable discussions. Authors would also like to thank to Dr. B. R. Mehta and Dr. Deepak Varandani, IIT Delhi, for AFM measurements. One of the Authors, Tanuj Dhawan, is thankful to Director, SSPL, for providing experimental facilities and a research fellowship.

REFERENCES

1. Liu HY, Hopkinson M, Groom K, Hogg RA, Mowbray DJ: Proc. SPIE Int. Soc. Opt. Eng.. 2008, **6909:** 690903. 10.1117/12.765735
2. Mi Z, Wu C, Yang J, Bhattacharya P: J. Vac. Sci. Technol. B Microelectron. Nanometer. Struct.. 2008, **26:** 1153. COI number [1:CAS:528:DC%2BD1cXms1Cgurc%3D]; Bibcode number [2008JVSTB..26.1153M] COI number

[1:CAS:528:DC%2BD1cXms1Cgurc%3D]; Bibcode number [2008JVSTB..26.1153M] 10.1116/1.2889386

3. Pal D, Walker J, Towe E: J. Vac. Sci. Technol. B Microelectron. Nanometer. Struct.. 2006, **24:** 1532. COI number [1:CAS:528:DC%2BD28XlvFWju7c%3D]; Bibcode number [2006JVSTB..24.1532P] COI number [1:CAS:528:DC%2BD28XlvFWju7c%3D]; Bibcode number [2006JVSTB..24.1532P] 10.1116/1.2190675

4. Kim E-T, Madhukara A, Ye Z, Campbell JC: Appl. Phys. Lett.. 2004, **84:** 3277. COI number [1:CAS:528:DC%2BD2cXjsVKls70%3D]; Bibcode number [2004ApPhL..84.3277K] COI number [1:CAS:528:DC%2BD2cXjsVKls70%3D]; Bibcode number [2004ApPhL..84.3277K] 10.1063/1.1719259

5. Park G, Shchekin OB, Huffaker DL, Deppe DG: IEEE Photon. Technol. Lett.. 2000, **12:** 230. Bibcode number [2000IPTL...12..230P] Bibcode number [2000IPTL...12..230P] 10.1109/68.826897

6. Grundmann M, Bimberg D: Jpn. J. Appl. Phys.. 1997, **36:** 4181. COI number [1:CAS:528:DyaK2sXkvFGgtr8%3D]; Bibcode number [1997JaJAP..36.4181G] COI number [1:CAS:528:DyaK2sXkvFGgtr8%3D]; Bibcode number [1997JaJAP..36.4181G] 10.1143/JJAP.36.4181V

7. Shchekin OB, Ahn J, Deppe DG: Electron. Lett.. 2002, **38:** 712. 10.1049/el:20020509

8. Kovsh AR, Maleev NA, Zhukov AE, Mikhrin SS, Vasil'ev AR, Shemyakov Yu M, Maximov MV, Livshits DA, Ustinov V, Alferov Zh I, Ledentsov NN, Bimberg D: Electron. Lett.. 2002, **38:** 1104. COI number [1:CAS:528:DC%2BD38XoslWjtLw%3D] COI number [1:CAS:528:DC%2BD38XoslWjtLw%3D] 10.1049/el:20020793

9. Ribbat Ch, Sellin RL, Kaiander I, Hopfer F, Ledentsov NN, Bimberg D, Kovsh AR, Ustinov VM, Zhukov AE, Maximov MV: Appl. Phys. Lett.. 2003, **82:** 952. COI number [1:CAS:528:DC%2BD3sXhtVGqtrg%3D]; Bibcode number

[2003ApPhL..82..952R] COI number [1:CAS:528:DC%2BD3sXhtVGqtrg%3D]; Bibcode number [2003ApPhL..82..952R] 10.1063/1.1533841

10. Aggarwal SK, Tyagi R, Singh M, Jain RK: Sol. Energy Mater. Sol. Cells. 1999, **59:** 19. 10.1016/S0927-0248(99)00027-6

11. D'Hondt M, Yu Z-Q, Depreter B, Sys C, Moerman I, Demeester P, Mijlemans P: J. Cryst. Growth. 1998, **195:** 655. 10.1016/S0022-0248(98)00652-6

12. Hjiri M, Hassen F, Maaref H: Mater. Sci. Eng. B. 2000, **69–70:** 514. 10.1016/S0921-5107(99)00248-2

13. Lee Y, Ahn E, Kim J, Moon P, Yang C, Yoon E, Lim H, Cheong H: Appl. Phys. Lett.. 2007, **90:** 033105. Bibcode number [2007ApPhL..90c3105L] Bibcode number [2007ApPhL..90c3105L] 10.1063/1.2432285

14. El-Emawy AA, Birudavolu S, Wong PS, Jiang YB, Xu H, Huang S, Huffaker DL: J. Appl. Phys.. 2003, **93:** 3529. COI number [1:CAS:528:DC%2BD3sXhvV2qs7o%3D]; Bibcode number [2003JAP....93.3529E] COI number [1:CAS:528:DC%2BD3sXhvV2qs7o%3D]; Bibcode number [2003JAP....93.3529E] 10.1063/1.1543647

15. Modak P, D'Hondt M, Mijlemans P, Moerman I, Van Daele P, Demeester P: J. Electron. Mater.. 2000, **29:** 80. COI number [1:CAS:528:DC%2BD3cXot1Kqtg%3D%3D]; Bibcode number [2000JEMat..29...80M] COI number [1:CAS:528:DC%2BD3cXot1Kqtg%3D%3D]; Bibcode number [2000JEMat..29...80M] 10.1007/s11664-000-0099-z

16. Knuuttila L, Kainu K, Sopanen M, Lipsanen H: J. Mater. Sci. Mater. Electron. 2003,**14**(5–7):349. 10.1023/A:1023992432393

17. Kroemer H: J. Cryst. Growth. 1987, **81:** 193. COI number [1:CAS:528:DyaL2sXhs1Knsrc%3D]; Bibcode number [1987JCrGr..81..193K] COI number [1:CAS:528:DyaL2sXhs1Knsrc%3D]; Bibcode number [1987JCrGr..81..193K] 10.1016/0022-0248(87)90391-5

18. Petroff PM: J. Vac. Sci. 1986, **4:** 874. COI number [1:CAS:528:DyaL28XlsVektbs%3D] COI number [1:CAS:528:DyaL28XlsVektbs%3D] 10.1116/1.583529

19. Li Y, Lazzarini L, Giling LJ, Salviati G: J. Appl. Phys.. 1994, **76:** 5748. COI number [1:CAS:528:DyaK2MXitFemsL0%3D]; Bibcode number [1994JAP....76.5748L] COI number [1:CAS:528:DyaK2MXitFemsL0%3D]; Bibcode number [1994JAP....76.5748L] 10.1063/1.358412

20. Lazzarini L, Nasi L, Salviati G, Fregonara CZ, Li Y, Giling LJ, Hardingham C, Holt DB: Micron. 2000, **31:** 217. 10.1016/S0968-4328(99)00086-4

21. Li W, Laaksonen S, Haapamaa J, Pessa M: J. Cryst. Growth. 2001, **227–228:** 104. 10.1016/S0022-0248(01)00641-8

22. Li Y, Salviati G, Bongers MMG, Lazzarini L, Nasi L, Giling LJ: J. Cryst. Growth. 1996, **163:** 195. COI number [1:CAS:528:DyaK28XjtlGrs7c%3D]; Bibcode number [1996JCrGr.163..195L] COI number [1:CAS:528:DyaK28XjtlGrs7c%3D]; Bibcode number [1996JCrGr.163..195L] 10.1016/0022-0248(95)00958-2

23. Tyagi R, Singh M, Thirumavalavan M, Srinivasan T, Aggarwal SK: J. Electron. Mater.. 2002,**31**(3):234. 10.1007/s11664-002-0212-6

24. Hudait MK, Krupanidhi SB: J. Appl. Phys.. 2001, **89:** 5972. COI number [1:CAS:528:DC%2BD3MXkt1WrtLs%3D]; Bibcode number [2001JAP....89.5972H] COI number [1:CAS:528:DC%2BD3MXkt1WrtLs%3D]; Bibcode number [2001JAP....89.5972H] 10.1063/1.1368870

25. Tan HH, Sears K, Mokkapati S, Fu L, Kim Y, McGowan P, Buda M, Jagdish C: IEEE J. Sel. Top. Quantum Electron.. 2006, **12:** 1242. COI number [1:CAS:528:DC%2BD2sXjsleg] COI number [1:CAS:528:DC%2BD2sXjsleg]

26. Wang ZM: Self-Assembled Quantum Dots. 2008. Springer Springer

10 CHAPTER

QUANTUM DOTS AS MULTIFUNCTIONAL MATERIALS FOR TUMOR IMAGING AND THERAPY

Longfei Liu, Qingqing Miao and Gaolin Liang *

CAS Key Laboratory of Soft Matter Chemistry, Department of Chemistry, University of Science and Technology of China, 96 Jinzhai Road, Hefei, Anhui 230026, China

ABSTRACT

The rapidly developing field of quantum dots (QDs) provides researchers with more options for imaging modalities and therapeutic strategies. In recent years, QDs were widely used as multifunctional materials for tumor imaging and therapy due to their characteristic properties such as semiconductive, zero-dimension and strong fluorescence. Nevertheless, there still exist the challenges of employing these properties of QDs for clinical diagnosis and therapy. Herein, we briefly review the development, properties and applications of QDs in tumor imaging and therapy. Future perspectives in these areas are also proposed as well.

Keywords: quantum dots; tumor imaging and therapy; semiconductor; fluorescence; probe

1. INTRODUCTION

1.1. Quantum Dots

The quantum dot (QD) is defined as an artificially structured system with the capacity to load electrons [1]. Its special physicochemical properties differentiate it from other naturally occurring biogenic and anthropogenic nanoparticles [2]. QDs are one type of nanoparticles (NPs) with three characteristic properties: semiconductors, zero-dimension, and strong fluorescence. Although colloidal semiconductor QDs are single crystals with diameters of a few nanometers, their sizes and shapes can be precisely controlled by the duration, temperature, and ligand molecules during the synthetic processes [3]. The well controlled synthetic process yields QDs with composition- and size-dependent absorption and emission. Generally, as the sizes of QDs are reduced, the electronic excitations shift to higher energies (i.e., shorter wavelengths) [4], such as CdSe QDs (shown in Figure 1) [5]. Besides, physical properties of QDs influence their fluorescence emissions. The fluorescence emission of colloidal QDs on the surface of a two-dimensional slab of photonic crystals will be enhanced due to a combination of high-intensity near fields with strong coherent scattering, which is related to leaky eigenmodes of the photonic crystal (In QDs' periodically modulated structures, an anomalous resonant phenomenon arises from periodic index modulation of the refractive index, which allows phase-matching of externally incident radiation into modes that can be re-radiated into free space). Owing to the fact that these modes possess finite lifetimes within such structures, they are called "leaky eigenmodes" [6]. Another example of the fluorescence enhancement of QDs is that when a sharp gold tip is brought within a few nanometers from a single QD cluster surface, the fluorescence of the QD in the vicinity of the tip increases about fourfold in magnitude [7].

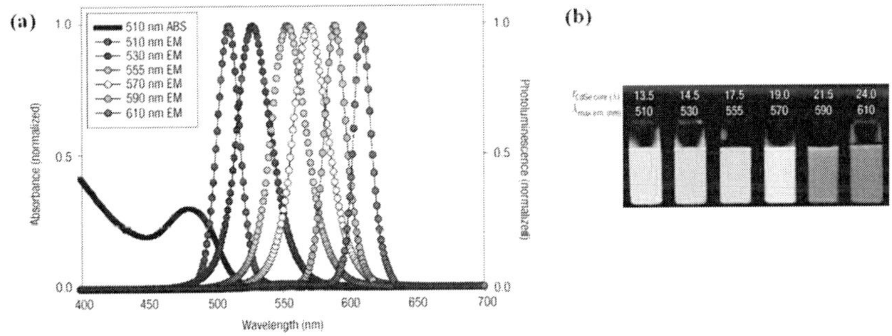

Figure 1. (a) Absorption and emission of six different quantum dot (QD) dispersions. The black line shows the absorption of the 510 nm emitting QDs. Note that at the wavelength of lowest absorption for the 510 nm QD, ~450 nm, the molar extinction coefficient is greater than that of rhodamine red at its absorption maxima (~150,000 vs. 129,000 M−1 cm−1); (b) Photo demonstrating the size-tunable fluorescence properties and spectral range of the six QD dispersions plotted in A vs. CdSe core size. All samples were excited at 365 nm with a UV source. For the 610 nm emitting QDs, this translates into a Stokes shift of ~250 nm. r = radius. Reprinted with permission from [5]. Copyright 2005 Nature Publishing Group.

The specific properties of QDs enable them with wide applications in chemistry, chemical biology and biomedicine. In the past decade, QDs have been broadly applied in fluorescence resonance energy transfer (FRET) analysis, gene technology, fluorescent labeling of cellular proteins, cell tracking, in vivo animal imaging and tumor biology investigation [8]. In this review, we will discuss the applications of QDs in tumor imaging and therapy.

1.2. Tumor Imaging

Tumor imaging is an imaging technology, which monitors changes of tumor cells at tissue, cellular, or sub-cellular levels. It helps people with qualitative and quantitative analyses of the biological processes of tumors in imaging aspects. The three predominant imaging modalities are optical imaging (e.g., fluorescence or non-fluorescence imaging), nuclear imaging (e.g., single photon emission computed tomography (SPECT) and positron emission tomography (PET)), and Magnetic Resonance Imaging (MRI). Figure 2 and Table 1 show the development of imaging techniques and their characteristics, respectively [9]. Compared with other imaging methods, QDs have advantages of broad light spectrum emission from visible to infrared due to their controllable sizes, bright and photostable fluorescence within a few nanometers, and good water-solubility [10]. However, QDs have inherent disadvantages such as high cytotoxicity (e.g., QDs of CdSe) [10].

As one of the optical imaging modalities, fluorescence refers to relatively-longer-wavelength light (especially visible light) emitted by certain molecule after it absorbing light at a particular wavelength [9]. The fluorophores are classified into two groups in terms of their origin: Endogenous fluorophores and exogenous fluorophores. For example, NADH (reduced form of nicotinamide-adenine dinucleotide) is one type of endogenous fluorophores, which can indicate the metabolic status of tumor because only its reduced form has fluorescence [11]. Exogenous fluorophores are usually organic compounds with fluorogenic motifs (e.g., near-infrared (NIR) fluorophores such as heptamethine cyanines containing benzoxazole or benzothiazole motifs [12].

As the two main modalities for nuclear imaging, SPECT and PET play important roles of imaging bone metastases in miscellaneous cancers including lung cancer, thyroid cancer, renal cancer, myeloma, and neuroendocrine cancers [13]. While SPECT uses one photon with lower energy to produce three-dimensional image of tracer distribution with multiplanar images, PET uses positron-emitting radiotracers and achieves image with higher spatial resolution than that of SPECT [13]. MRI is also frequently applied in tumor imaging using dotarem or magnevist as T1 contrast media or resovist as T2 contrast medium [14]. To date, multimodalities of imaging such as PET-CT, PET-MRI have emerged and been used for the acquisition of images with higher accuracy [9].

QDs plays an increasingly important role in tumor imaging, especially near-infrared (NIR, 700–900 nm) imaging. NIR fluorescence imaging of tumor is expected to have a major impact in biomedical imaging because in the NIR region the absorbance spectra for all the biomolecules in tumor reach their minima, which provides a clear window for in vivo optical imaging of tumor [15]. In 2010, Gao et al. reported that QD800-MPA (a NIR non-cadmium QDs coated with mercaptopropionic acid with an emission wavelength of about 800 nm) had high tumor uptake and excellent contrast of tumor to surrounding tissues due to the enhanced permeability and retention (EPR) effect of this kind of ultrasmall nanoparticles [16].

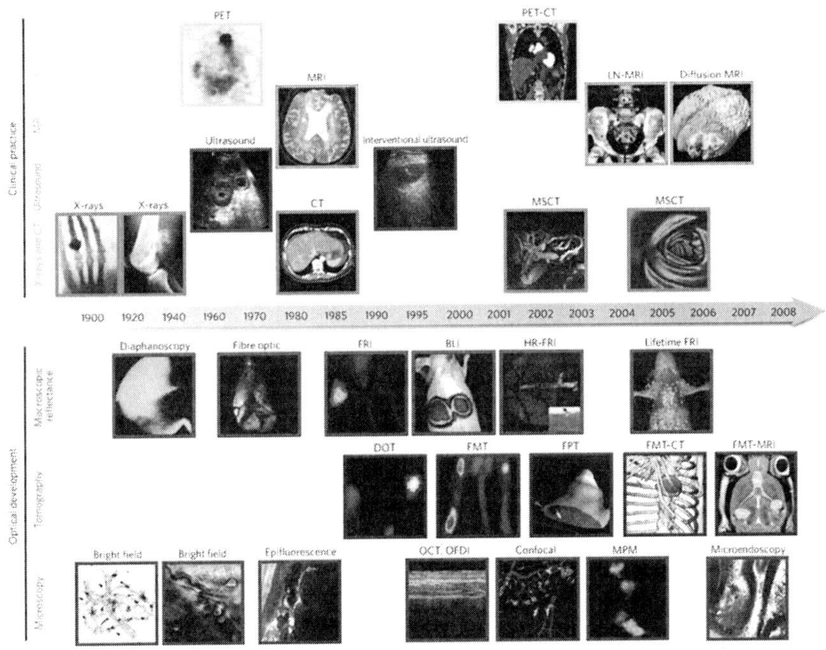

Figure 2. Imaging technologies used in oncology. Many macroscopic imaging technologies (shown above the timeline) are in routine clinical use, and there have been huge advances in their capabilities to obtain anatomical and physiological information since the beginning of the twentieth century. Shown are some examples of bones (X-rays), soft tissue (ultrasound, MRI and CT rows), three-dimensional organs (CT and MRI rows) and physiological imaging (MRI and PET rows). Microscopic and other intravital optical techniques (shown below the timeline) have developed over the past decade and now allow studies of genetic, molecular and cellular events in vivo. Shown are surface-weighted, whole-mouse, two-dimensional techniques (macroscopic reflectance row); tomographic

three-dimensional techniques, often in combination with other anatomical modalities (tomography row); and intravital microscopy techniques (microscopy row). The timeline is approximate and is not to scale. Here BLI, bioluminescence imaging; CT, computed tomography; DOT, diffuse optical tomography; FMT, fluorescence-mediated tomography; FPT, fluorescence protein tomography; FRI, fluorescence reflectance imaging; HR-FRI, high-resolution FRI; LN-MRI, lymphotropic nanoparticle-enhanced MRI; MPM, multiphoton microscopy; MRI, magnetic resonance imaging; MSCT, multislice CT; OCT, optical coherence tomography; OFDI, optical frequency-domain imaging; PET, positron-emission tomography. Reprinted with permission from [9]. Copyright 2008 Nature Publishing Group.

Table 1. Overview of imaging systems. Reproduced with permission from [9]. Copyright 2008 Nature Publishing Group.

Technique	Resolution [*]	Depth	Time [†]	Quantitative [‡]	Multi-channel	Imaging agents	Target	Cost [§]	Main small-animal use	Clinical use
MRI	10–100 μm	No limit	Minutes to hours	Yes	No	Paramagnetic chelates, magnetic particles	Anatomical, physiological, molecular	$$$	Versatile imaging modality with high soft-tissue contrast	Yes
CT	50 μm	No limit	Minutes	Yes	No	Iodinated molecules	Anatomical, physiological	$$	Imaging lungs and bone	Yes
Ultrasound	50 μm	cm	Seconds to minutes	Yes	No	Microbubbles	Anatomical, physiological	$$	Vascular and interventional imaging [‖]	Yes
PET	1–2 mm	No limit	Minutes to hours	Yes	No	^{18}F-, ^{64}Cu- or ^{11}C-labelled compounds	Physiological, molecular	$$$	Versatile imaging modality with many tracers	Yes
SPECT	1–2 mm	No limit	Minutes to hours	Yes	No	^{99m}Tc- or ^{111}In-labelled compounds	Physiological, molecular	$$	Imaging labelled antibodies, proteins and peptides	Yes
Fluorescence reflectance imaging	2–3 mm	<1 cm	Seconds to minutes	No	Yes	Photoproteins, fluorochromes	Physiological, molecular	$	Rapid screening of molecular events in surface-based disease	Yes
FMT	1 mm	<10 cm	Minutes to hours	Yes	Yes	Near-infrared fluorochromes	Physiological, molecular	$$	Quantitative imaging of fluorochrome reporters	In development
Bioluminescence imaging	Several mm	cm	Minutes	No	Yes	Luciferins	Molecular	$$	Gene expression, cell and bacterium tracking	No
Intravital microscopy [¶]	1 μm	<400–800 μm	Seconds to hours	No	Yes	Photoproteins, fluorochromes	Anatomical, physiological, molecular	$$$	All of the above at higher resolutions but limited depths and coverage	In development [#]

[*] For high-resolution, small-animal imaging systems. (Clinical imaging systems differ.); [†] Time for image acquisition; [‡] Quantitative here means inherently quantitative. All approaches allow relative quantification; [§] Cost is based on purchase price of imaging systems in the United States: $, <US$100,000; $$, US$100,000–300,000; $$$, >US$300,000; [‖] Interventional means used for interventional procedures such as biopsies or injection of cells under ultrasound guidance; [¶] Laser-scanning confocal or multiphoton microscopy; [#] For microendoscopy and skin imaging.

1.3. Tumor Therapy

Millions of people die from cancer every year, especially from lung cancer. Even though no existing method can defeat cancer, tumor therapies such as surgery, radiotherapy, chemotherapy, and photodynamic therapy (PDT) are developing dramatically nowadays.

Surgery has relatively good effects for benign tumors and precancerous tumors. Advances in radiotherapy, such as intensity-modulated radiation therapy (IMRT), provide the capability of delivering a highly conformal distribution of radiative dose to a static, complex targeting volume [17]. Chemotherapy is widely adopted for tumor therapy and hundreds of anticancer drugs are used clinically. Typical anticancer drugs include plant extractives (e.g., taxol [18]), heavy metal complexes (e.g., cisplatin [19]), bioreductive drugs (e.g., tirapazamine, 3-amino-1,2,4-benzotriazine 1,4-dioxide or TPZ [20]), and traditional Chinese medicine [21]. In bcl-2-positive cancer cells (cancer cells expressing bcl-2), taxol induces the phosphorylation of bcl-2 and programmed cell death thereafter [22]. Cisplatin, one of the most widely used anticancer drugs, can bind with DNA to form cis-DDP/DNA adducts which induce DNA-damage and cell apoptosis [19]. Tirapazamine (TPZ), a leading bioreductive drug with selective cytotoxicity to hypoxic cells in tumor, damages the DNA inside cell with the reactive oxygen species (ROS) with its reduction product under the action of reductases in cells [20].

PDT has been an innovative and attractive modality for treatment of small and superficial tumors since the end of the last century [23]. After absorption of light with certain wavelength, sensitizers can induce the necrosis of tumors [23]. Some examples of QDs for PDT of tumors are discussed in the following parts.

2. QDS FOR TUMOR IMAGING

QDs, tiny light-emitting particles on nanometer scale, are new type of fluorescent probes for molecular and cellular imaging. Compared with organic dyes and fluorescent proteins, QDs have unique optical and electronic properties in cellular imaging: Wavelength-tunable emission, improved brightness of signal, resistance against photobleaching, etc. [24]. Such preponderant optical properties were not realized until the QD-based probes are equipped with war heads targeting tumor. Xingyong Wu and co-workers synthesized immunofluorescent probes by conjugating the QDs with streptavidin or IgGs (immunoglobulin Gs). Using the conjugates, they conducted comprehensive investigations on cell imaging at the targets of interest including cell surface receptors, cytoskeleton components, and nuclear antigens [25]. Up to date, QDs have been rapidly developed in tumor imaging, such as imaging tumor vasculature [15] and sentinel lymph node [26].

2.1. QDs for Imaging Membrane Receptors (Surface)

Metastases, which are responsible for most cancer deaths rather than those of primary tumors, spread tumor cells from a primary site to new distant organs [27]. Changes of membrane morphology or dynamics of membrane protein in cancer cells for cellular fluidity are critical for cancer metastasis [28]. Therefore, QDs modified with targeting ligands offer a good opportunity to track the changes of related membrane receptors. Soonhag Kim and co-workers designed a series of aptamers conjugating with different QDs to image the proteins on the membranes of cancer cells [29]. They chose three different QDs with distinct emission wavelengths of 605, 655 and 705 nm to respectively conjugate with three cancer-related aptamers-AS1411, TTA1, and MUC-1 [29]. AS1411 is an aptamer that binds to the nucleolin in the plasma membranes of cancer cells [30,31]. TTA1, which is expressed during the tissue remodeling processes including angiogenesis, inflammation, and tumor growth, binds to the extracellular matrix protein tenascin-C of cancer cells [32]. MUC-1 targets mucin, which is highly expressed by the majorities of human adenocarcinomas [33,34]. Confocal microscopic cell images were obtained with the receptors in different cells being successfully labeled with respective QD-conjugated aptamers (shown in Figure 3) [29]. Gonda et al. [28] also labeled a metastasis-promoting factor on the cell membrane called protease-activated receptor 1 (PAR1) with QD-conjugated anti-PAR1 antibody. By tracking the fluorescence of QDs, they photographed four stages of metastasis: cancer cells far from blood vessels in tumor, near the vessel, in the bloodstream, and adherent to the inner vascular surface in the normal tissues near tumor. With this, they successfully showed the dynamics of PAR1 movement in the whole process [28].

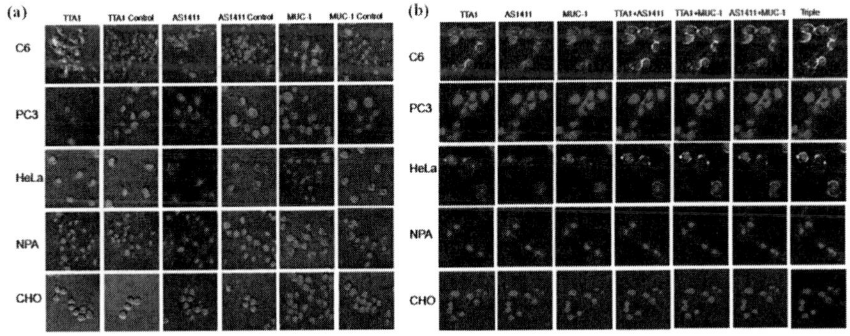

Figure 3. Confocal microscopy imaging of cells treated with QD-aptamer conjugates. (a) A single incubation of QD605-TTA1, QD655-AS1411, or QD705-MUC-1 was applied to PC-3, HeLa, CHO, C6, and NPA cells, and confocal images were obtained. Each image was compared with the corresponding QD-control aptamers (column 1: QD-TTA1, column 2: QD-TTA1 control, column 3: QD-AS1411, column 4: QD-AS1411 control, column 5: QD-MUC-1, column 6: QDMUC- 1control); (b) Multiplex imaging of cancer cells treated simultaneously with three different types of QD-conjugated aptamers. Single images for QD-TTA1 (605 nm, light green, column 1), QD-AS1411 (655 nm, red, column 2), and QD-MUC-1(705 nm, violet, column3), dual images for QD-AS1411 and QD-TTA1 (column 4, yellow for co-localization), QD-TTA1 and QD-MUC-1 (column 5, light green for co-localization), and QD-AS1411 and QD-MUC-1 (column 6, violet for co-localization), and a triple image for QD-AS1411, QD-TTA1, and QD-MUC-1 (column 7, white for co-localization) were acquired from PC-3, HeLa, CHO, C6, and NPA cells. All figures are merged with the 40, 6-diamidino-2-phenylindole (DAPI) image (nucleus staining, 460 nm) and cellular morphology. Reprinted with permission from [29]. Copyright 2009 John Wiley and Sons.

2.2. QDs for Imaging Cytoskeleton Components (Intracellular)

The cellular cytoskeleton, involved in many fundamental processes (e.g., locomotion and cytokinesis) of the cell, consists of actin filaments, microtubules and intermediate filaments [35]. Tumor cells

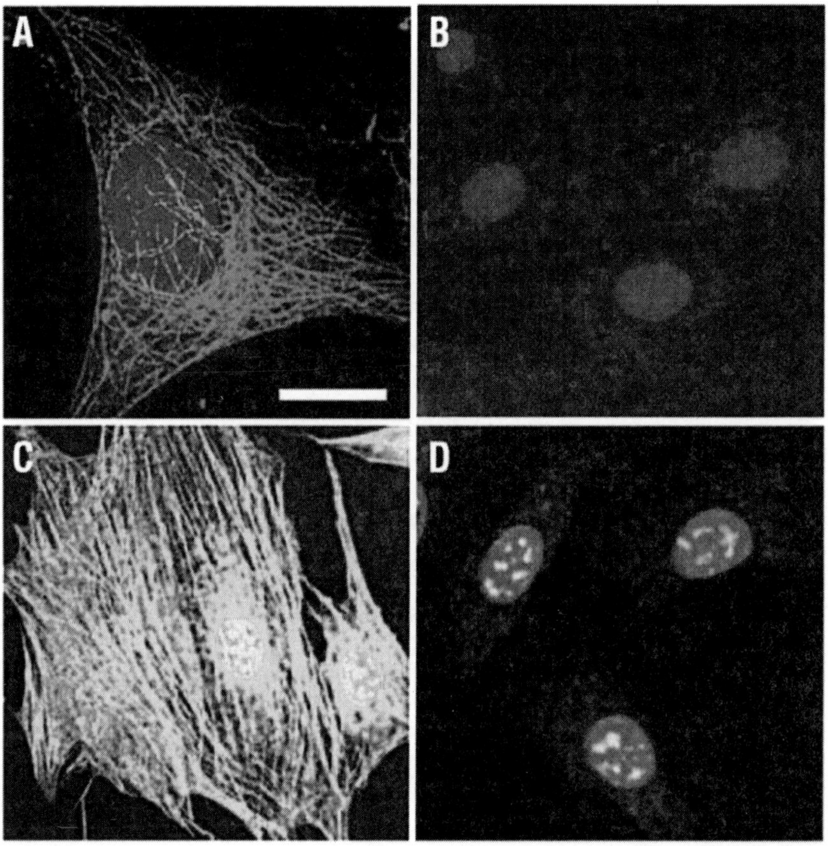

Figure 4. Staining of cytoskeleton fibers in 3T3 mouse fibroblast cells with QD-streptavidin. (A) Microtubules were labeled with monoclonal anti-α-tubulin antibody, biotinylated anti-mouse IgG and QD 630-streptavidin (red); (B) Control for (A) without primary antibody; (C) Actin filaments were stained with biotinylated phalloidin and QD 535-streptavidin (green); (D) Control for (C) without biotin-phalloidin. The nuclei were counterstained with Hoechst 33342 blue dye. Filter sets ex. 480 ± 20 nm/em. 535 ± 10 nm and ex. 560 ± 27.5 nm/em. 635 ± 10 nm were used to observe signals of QD 535 and QD 630, respectively. Scale bar, 10 μm for (A), 24 μm for (B) through (D). Reprinted with permission from [25]. Copyright 2003 Nature Publishing Group.

are in endless division, which is related to the movement of actin filaments and microtubules [36,37]. Therefore, imaging the movement of actin filaments and microtubules in tumor cells is important for tumor imaging.

In the end of the last century, Bruchez et al. labeled the F-actin filaments with red nanocrystal probes conjugated with biotin [38]. Compared with conventional dye molecules, the nanocrystal-labeled samples showed advantages of neglectable photobleaching [38]. Wu et al. used QD 630-streptavidin (red) and QD 535-streptavidin (green) to stain microtubules and actin filaments, respectively [25]. The results indicated the QD-based probes could be bright enough and specific enough for effectively labeling fine cellular structures, and have a better performance over other probes reported (shown in Figure 4) [25]. In 2008, Higuchi et al. reported new photostable, bright QDs conjugated with anti-tubulin antibody, which could bind to microtubules and trace the dynamic movement of microtubules in living cancer cells [39].

2.3. QDs for Imaging Nuclear Antigens (Intranuclear)

Tumor cells have some specific over-expressed nuclear antigens relating to their endless proliferation, such as PCNA (proliferating cell nuclear antigen) [40]. It was reported that QDs coated with urea or acetate groups might stain the nucleus [38]. Tang et al. used CdSe/ZnS QDs conjugated with anti-human PCNA antibody to label PCNA (proliferating cell nuclear antigens) in breast cancer tissues (shown in Figure 5) [40]. Xingyong Wu used QD 630-streptavidin to label the nuclei of SK-BR-3 cells successfully [25]. Labeling nuclear antigens in tumor cells with QD-conjugated bioprobes offers people with useful and reliable information for biomedical analysis and cancer diagnosis.

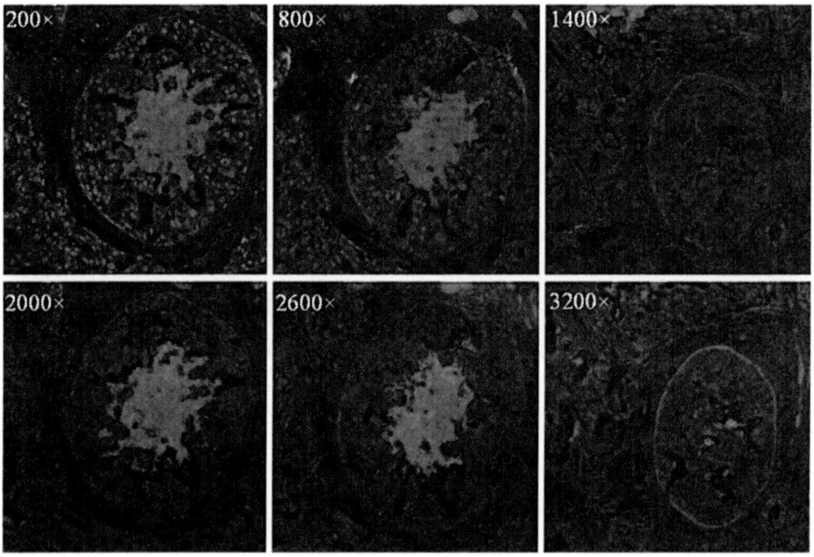

Figure 5. Fluorescence images of breast cancer tissues labeled with CdSe/ZnS QDs. PCNA was stained red with QDs modified with the antibody. The original QD-SA concentration was about 1 μmol/L and was diluted 200–3200× before staining operations. Reprinted with permission from [40]. Copyright 2010 Society for Applied Spectroscopy.

2.4. QDs for Imaging Tumor Neovasculature (a Special Example of Tumor Imaging)

Newly formed/forming blood vessels express αvβ3 integrin, which specifically binds to arginine-glycine-aspartic (RGD) peptides. The αvβ3 integrin receptor plays an important role in tumor metastasis and tumor-induced angiogenesis, making it possible for RGD-conjugated QDs to image tumor neovasculature [41]. Employing this, Sanjiv Sam Gambhir and co-workers designed RGD-QDs for real-time intravital imaging of luminal endothelium in mouse tumor neovasculature [41]. The peptides for QDs conjugation are cyclo(RGDfC) and cyclo(RADfC). In contrast to the controls, RGD-QDs specifically bound to tumor vessel endothelium and exhibited better performance than organic dyes (shown in Figure 6). Importantly, this work of real-time imaging tumor neovasculature was performed in living subjects with an intravital microscopy, which opens the door of in vivo tumor imaging with QDs.

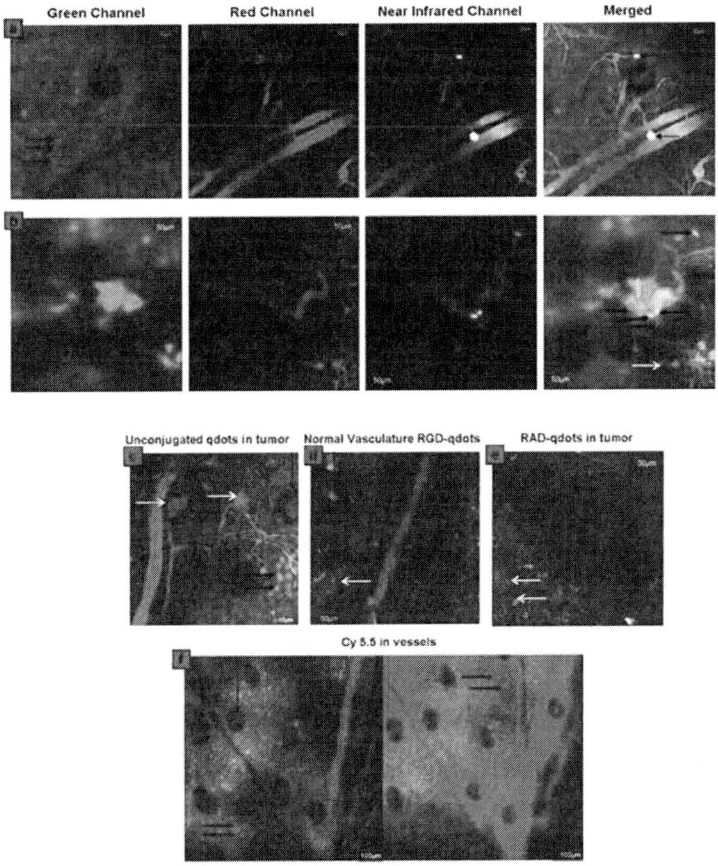

Figure 6. Direct visualization of binding of RGD-QDs to tumor vessel endothelium and controls. (a) Panel displays different output channels of the identical imaging plane along the row with scale bars. In the green channel, individual EGFP-expressing cancer cells are visible (marked by thick horizontal blue arrows; vertical blue arrow points to a hair follicle), while the red channel outlines the tumor's vasculature via injection of Angiosense dye. The NIR channel shows intravascularly administered QDs, which remain in the vessels (i.e., they do not extravasate). Binding events are visible by reference to bright white signal. These are demarcated by arrows in

the rightmost merged image, in which all three channels have been overlaid; (b) Displays the same as (a) in a different mouse, except that six times the RGD-QDs dose has been injected. Individual cells are not generally visible. Six binding events are observed in this FOV, as marked by arrows in the merged image at right. White arrows in the bottom merged image designate areas of tissue autofluorescence. Typical images of no binding in each control condition are shown in (c–f). Tumor neovasculature containing unconjugated QDs (c), normal vasculature containing RGD-QDs (d), and tumor neovasculature containing RAD-QDs (e). (f) Tumor vasculature shortly after Cy5.5 injection (left) and ~20 min post-Cy5.5 injection (right). Individual cancer cells are visible before (left) and after dye extravasates (right, dyed red). Also see movie S6 in Supporting Information. Horizontal white arrows indicate tissue autofluorescence, vertical blue arrows denote hair follicles (which generally display autofluorescence in their center), and thick horizontal blue arrows indicate individual cancer cells. Reprinted with permission from [41]. Copyright 2008 American Chemical Society.

3. QDS FOR TUMOR THERAPY

Although relatively fewer researches on QDs for tumor therapy were reported, it is conceivable that QDs have the potentialities for tumor therapy due to their large surface areas available for the modification of functional groups or therapeutic agents such as anti-cancer drugs [42] and PDT photosensitizers (PS) [43]. Moreover, QDs themselves can also functionalize as PDT photosensitizers for tumor therapy [44]. Herein, QDs as PDT photosensitizers and anti-cancer drug-QD complexes for cancer therapy are reviewed as following. Barberi-Heyo et al. established that QDs conjugated with folic acid (FA) could be used as PS for PDT of cancer [45]. They conjugated CdTe(S)-type QDs with FA which is an optimal targeting ligand for selectively delivering the attached therapeutic agents (herein QDs as PS) to cancer tissues. 3-(4,5-dimethylthiazol-2-yl) 2,5 diphenyl tetrazolium bromide (MTT) assay indicated that the survival rate of KB (human head and neck carcinoma cell line) cells incubated with FA-conjugated QDs decreased as the irradiation time or intensity increased (shown in Figure 7). The results demonstrated that CdTe(S)-type QDs had photosensitizing properties, which could be used to promote PDT effect. In their study, they mentioned that the concentration of QDs should be inferior to 10 nM and the incubation time less than 8 hour to avoid the intrinsic cytotoxicity of QDs without light irradiation.

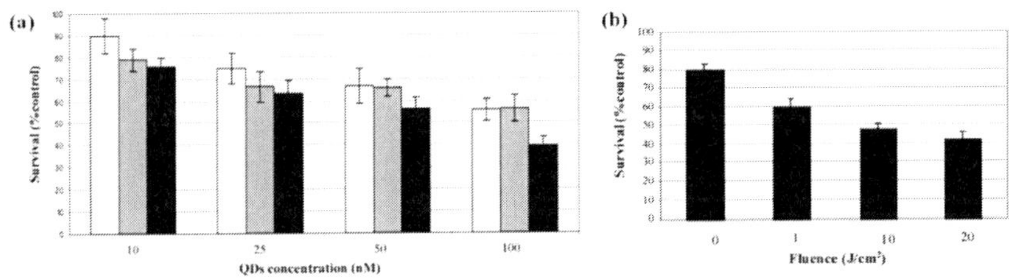

Figure 7. (a) Measurement of PDT sensitivity of KB cells treated with FA-conjugated QD 4. Cells were exposed to QDs in a concentration range of 10–100 nM for 1 (light gray), 3 (dark grey) and 6 h (black); (b) Survival curves obtained for cells incubated with QDs at 5 nM for 3 h incubation before irradiation to increasing doses of light from 1 to 20 J cm−2. Measurement of PDT sensitivity for the QDs were obtained by MTT test (data points show the mean ± s.d., n = 6). * $P < 0.05$ vs. previous fluence dose. Reprinted with permission from [45]. Copyright 2011 Royal Society of Chemistry.

Taking advantage of QDs' superior physical properties, PS-QDs conjugates can be excited with a wide range of wavelengths and avoid the PS to absorb light in the mean time (Figure 8 explains how QDs assist photosensitizers with producing singlet oxygen) [43]. Clemens Burda and co-workers used QD-based fluorescence resonance energy transfer (FRET) to facilitate the excitation of a PDT photosensitizer to generate reactive 1O2 species for PDT [46]. The results demonstrated that CdSe QDs could be used to sensitize either phthalocyanines (a family of PDT agents) such as Pc4 via FRET mechanism or itself as PS via a triplet energy transfer (TET) mechanism to produce 1O2 species for PDT (shown in Figure 9).

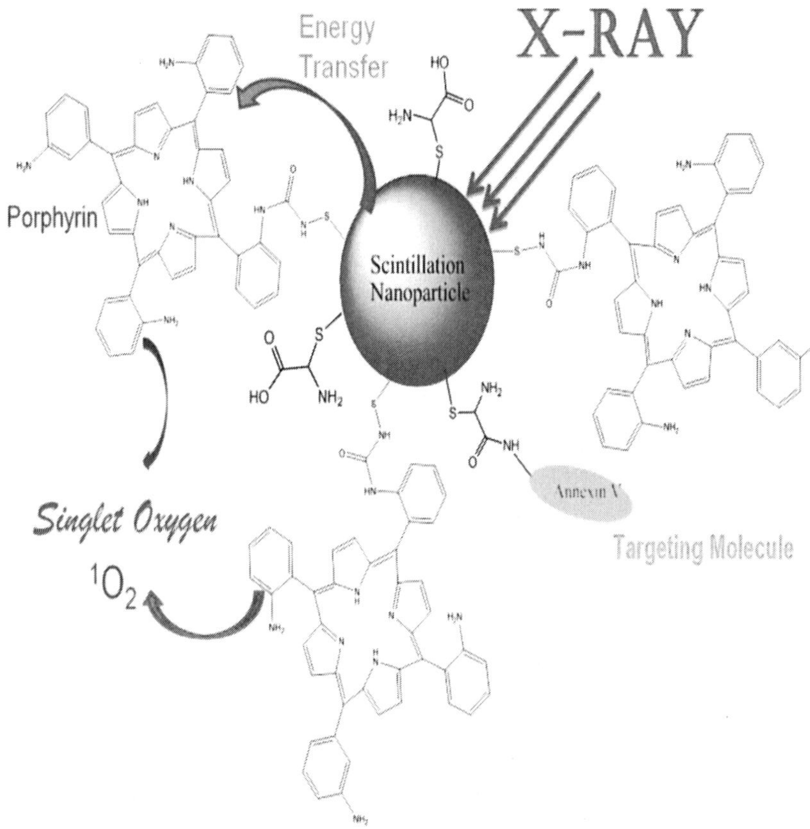

Figure 8. Schematic presentation of the nanoparticle-based X-ray-induced PDT. Under ionizing radiation a nanoparticle starts to scintillate transferring its energy into a conjugated porphyrin molecule, which then generates singlet oxygen necessary to produce photosensitizing effect. This methodology will help to treat nodular and deeper tumors due to higher penetrating capacity of X-rays and gamma rays compared to that of visible light commonly used in PDT. Reprinted with permission from [43]. Copyright 2008 Elsevier.

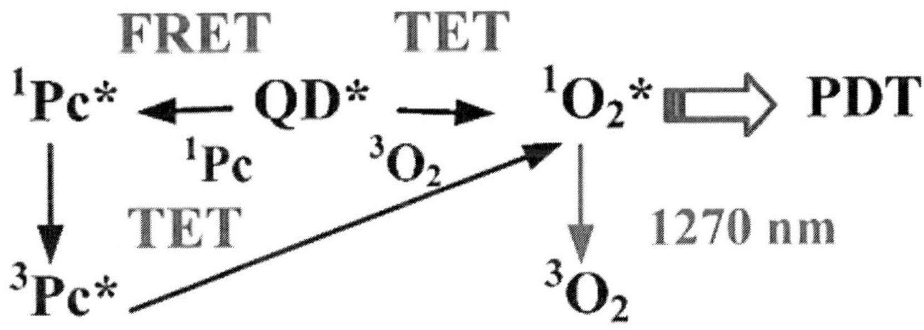

Figure 9. Schematics of the 1O2 generation in QD-based PDT systems. Reprinted with permission from [46]. Copyright 2003 American Chemical Society.

4. MULTIFUNCTIONAL QDS FOR SYNCHRONOUS TUMOR IMAGING AND THERAPY

QDs have exhibited specific advantages in tumor imaging and tumor therapy. Large surface area of QDs enables them to be conjugated with different agents including targeting molecules, therapeutic chemicals, and imaging substances. Obviously, the multifunctional QDs for synchronous tumor imaging and therapy will be much more attractive and important. Sangyong Jon and co-workers reported such a multifunctional QD-aptamer (Apt)-doxorubicin (Dox) conjugate [QD-Apt(Dox)] for cancer-targeted imaging, therapy, and sensing [47]. The conjugate consists of three components: QDs, which functionalize as fluorescent agents; RNA aptamers covalently attached to the surface of QD, which serve a dual functions as targeting molecules and as drug carrying vehicles; Dox, which is a therapeutic agent for tumor cells as well as a fluorescent agent (Dox has fluorescence itself [48]). This conjugate keeps fluorescence-off state through a Bi-FRET mechanism when Dox links to QD. Fluorescence of both QD and Dox will be turned on after Dox being released from the QD-conjugate. The results indicated that QD-Apt(Dox) could differentially bind to prostate specific membrane antigen(PSMA)-expressing LNCaP cells instead of the PSMA-negative

PC3 prostate adenocarcinoma cell lines due to the aptamer selectively binding to PSMA. Fluorescence microscopic cell imaging indicated that Dox was released from QD-conjugate 1.5 h after endocytosis and the targeting cells were stained with both Dox and QD (shown in Figure 10). MTT assay indicated that the QD-Apt(Dox) has LNCaP cell-targeted therapeutic ability (shown in Figure 11).

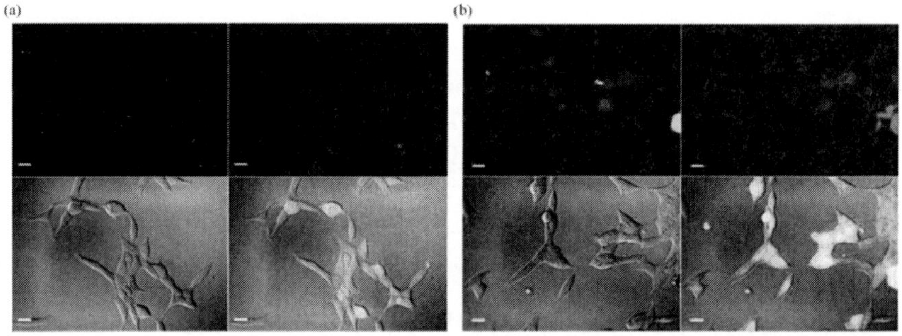

Figure 10. Confocal laser scanning microscopy images of PSMA expressing LNCaP cells after incubation with 100 nM QD-Apt-(Dox) conjugates for 0.5 h at 37 °C, washed two times with PBS buffer, and further incubated at 37 °C for (a) 0 h and (b) 1.5 h. Dox and QD are shown in red and green, respectively, and the lower right images of each panel represents the overlay of Dox and QD fluorescent. The scale bar is 20 μm. Reprinted with permission from [47]. Copyright 2007 American Chemical Society.

In summary, this conjugate can be used to detect cancer cells at a single cell level. It exhibits specificity and sensitivity to LNCaP cells for sensing, imaging, and therapy.

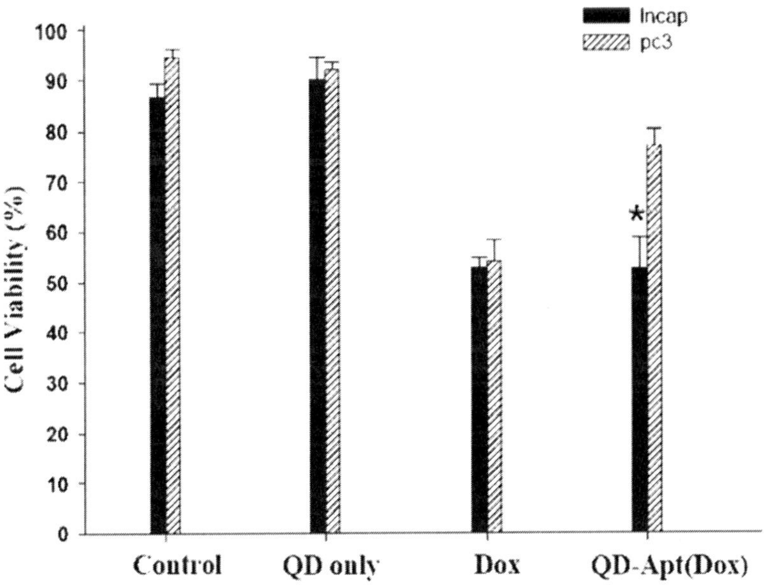

Figure 11. Growth inhibition assay (MTT). Prostate cancer cell lines, LNCaP (PSMA+) and PC3 (PSMA−), were incubated with QD alone (1.6 μM), Dox along (5 μM), or QD-Apt(Dox) conjugates (1.6 μM), for 3 h, and the cells were washed and further incubated for 24 h prior to measurement of cell viability. Asterisk indicates significant differences between LNCaP and PC3 cells, ($P < 0.005$, $n = 3$). Reprinted with permission from [47]. Copyright 2007 American Chemical Society.

5. CONCLUSIONS

Quantum Dots, as one type of multifunctional materials, have shown promising advantages in tumor imaging and therapy due to their specific physicochemical properties. Nevertheless, they still have some non-neglectable limitations such as increased sizes after coating [49] and the cytotoxicities introduced [2,50]. As we reviewed above, some QDs have impressive effect of imaging tumor neovasculature, which is however too late for clinical diagnosis of cancer development. These call for the development of new types of

QDs for the detection of important biomarkers (e.g., furin) of cancers at early stages [51]. Development of QDs in the future will not be limited to tumor imaging or therapy, but could be a combination of two or multiple functions. We envision that QDs will become one type of promising material for real-time tumor-targeted imaging and therapy in the future.

ACKNOWLEDGEMENTS

This work was supported by the National Natural Science Foundation of China (21175122, 91127036), the Fundamental Research Funds for Central Universities (WK2060190018), and Anhui Provincial Natural Science Foundation (1108085J17).

REFERENCES

1. Hanson, R.; Kouwenhoven, L.P.; Petta, J.R.; Tarucha, S.; Vandersypen, L.M.K. Spins in few-electron quantum dots. Rev. Mod. Phys. 2007, 79, 1217–1265.
2. Hardman, R. A toxicologic review of quantum dots: Toxicity depends on physicochemical and environmental factors. Environ. Health Perspect. 2006, 114, 165–172.
3. Michalet, X.; Pinaud, F.F.; Bentolila, L.A.; Tsay, J.M.; Doose, S.; Li, J.J.; Sundaresan, G.; Wu, A.M.; Gambhir, S.S.; Weiss, S. Quantum dots for live cells, in vivo imaging, and diagnostics. Science 2005, 307, 538–544.
4. Samir, T.M.; Mansour, M.M.H.; Kazmierczak, S.C.; Azzazy, H.M.E. Quantum dots: Heralding a brighter future for clinical diagnostics. Nanomedicine 2012, 7, 1755–1769.
5. Medintz, I.L.; Uyeda, H.T.; Goldman, E.R.; Mattoussi, H. Quantum dot bioconjugates for imaging, labelling and sensing. Nat. Mater. 2005, 4, 435–446.

6. Ganesh, N.; Zhang, W.; Mathias, P.C.; Chow, E.; Soares, J.; Malyarchuk, V.; Smith, A.D.; Cunningham, B.T. Enhanced fluorescence emission from quantum dots on a photonic crystal surface. Nat. Nanotechnol. 2007, 2, 515–520.

7. Huang, F.M.; Festy, F.; Richards, D. Tip-enhanced fluorescence imaging of quantum dots. Appl. Phys. Lett. 2005, 87, 183101:1–183101:3.

8. Jamieson, T.; Bakhshi, R.; Petrova, D.; Pocock, R.; Imani, M.; Seifalian, A.M. Biological applications of quantum dots. Biomaterials 2007, 28, 4717–4732.

9. Weissleder, R.; Pittet, M.J. Imaging in the era of molecular oncology. Nature 2008, 452, 580–589.

10. Ghasemi, Y.; Peymani, P.; Afifi, S. Quantum dot: Magic nanoparticle for imaging, detection and targeting. Acta Biomed. 2009, 80, 156–165.

11. Stringari, C.; Nourse, J.L.; Flanagan, L.A.; Gratton, E. Phasor fluorescence lifetime microscopy of free and protein-bound nadh reveals neural stem cell differentiation potential. PLoS One 2012, 7, e48014.

12. Frangioni, J.V. In vivo near-infrared fluorescence imaging. Curr. Opin. Chem. Biol. 2003, 7, 626–634.

13. Chua, S.; Gnanasegaran, G.; Cook, G.J.R. Miscellaneous cancers (lung, thyroid, renal cancer, myeloma, and neuroendocrine tumors): Role of spect and pet in imaging bone metastases. Semin. Nucl. Med. 2009, 39, 416–430.

14. Barrett, T.; Kobayashi, H.; Brechbiel, M.; Choyke, P.L. Macromolecular mri contrast agents for imaging tumor angiogenesis. Eur. J. Radiol. 2006, 60, 353–366.

15. Cai, W.B.; Shin, D.W.; Chen, K.; Gheysens, O.; Cao, Q.Z.; Wang, S.X.; Gambhir, S.S.; Chen, X.Y. Peptide-labeled near-infrared quantum dots for imaging tumor vasculature in living subjects. Nano Lett. 2006, 6, 669–676.

16. Gao, J.H.; Chen, K.; Xie, R.G.; Xie, J.; Lee, S.; Cheng, Z.; Peng, X.G.; Chen, X.Y. Ultrasmall near-infrared non-cadmium

quantum dots for in vivo tumor imaging. Small 2010, 6, 256–261.

17. Jiang, S.B. Radiotherapy of mobile tumors. Semin. Radiat. Oncol. 2006, 16, 239–248.

18. Guerra-Bubb, J.; Croteau, R.; Williams, R.M. The early stages of taxol biosynthesis: An interim report on the synthesis and identification of early pathway metabolites. Nat. Prod. Rep. 2012, 29, 683–696.

19. Galluzzi, L.; Senovilla, L.; Vitale, I.; Michels, J.; Martins, I.; Kepp, O.; Castedo, M.; Kroemer, G. Molecular mechanisms of cisplatin resistance. Oncogene 2012, 31, 1869–1883.

20. Anderson, R.F.; Shinde, S.S.; Hay, M.P.; Denny, W.A. Potentiation of the cytotoxicity of the anticancer agent tirapazamine by benzotriazine n-oxides: The role of redox equilibria. J. Am. Chem. Soc. 2006, 128, 245–249.

21. Efferth, T.; Li, P.C.H.; Konkimalla, V.S.B.; Kaina, B. From traditional chinese medicine to rational cancer therapy. Trends Mol. Med. 2007, 13, 353–361. Ping, S.Y.; Hour, T.C.; Lin, S.R.; Yu, D.S. Taxol synergizes with antioxidants in inhibiting hormal refractory prostate cancer cell growth. Urol. Oncol. Semin. Orig. Investig. 2010, 28, 170–179.

22. Robertson, C.A.; Evans, D.H.; Abraharnse, H. Photodynamic therapy (PDT): A short review on cellular mechanisms and cancer research applications for pdt. J. Photochem. Photobiol. B Biol. 2009, 96, 1–8.

23. Gao, X.H.; Yang, L.L.; Petros, J.A.; Marshal, F.F.; Simons, J.W.; Nie, S.M. In vivo molecular and cellular imaging with quantum dots. Curr. Opin. Biotechnol. 2005, 16, 63–72.

24. Wu, X.Y.; Liu, H.J.; Liu, J.Q.; Haley, K.N.; Treadway, J.A.; Larson, J.P.; Ge, N.F.; Peale, F.; Bruchez, M.P. Immunofluorescent labeling of cancer marker Her2 and other cellular targets with semiconductor quantum dots. Nat. Biotechnol. 2003, 21, 41–46.

25. Ballou, B.; Ernst, L.A.; Andreko, S.; Harper, T.; Fitzpatrick, J.A.J.; Waggoner, A.S.; Bruchez, M.P. Sentinel lymph node

imaging using quantum dots in mouse tumor models. Bioconjug. Chem. 2007, 18, 389–396.

26. Chambers, A.F.; Groom, A.C.; MacDonald, I.C. Dissemination and growth of cancer cells in metastatic sites. Nat. Rev. Cancer 2002, 2, 563–572.

27. Gonda, K.; Watanabe, T.M.; Ohuchi, N.; Higuchi, H. In vivo nano-imaging of membrane dynamics in metastatic tumor cells using quantum dots. J. Biol. Chem. 2010, 285, 2750–2757.

28. Kang, W.J.; Chae, J.R.; Cho, Y.L.; Lee, J.D.; Kim, S. Multiplex imaging of single tumor cells using quantum-dot-conjugated aptamers. Small 2009, 5, 2519–2522.

29. Christian, S.; Pilch, J.; Akerman, M.E.; Porkka, K.; Laakkonen, P.; Ruoslahti, E. Nucleolin expressed at the cell surface is a marker of endothelial cells in angiogenic blood vessels. J. Cell Biol. 2003, 163, 871–878.

30. Teng, Y.; Girvan, A.C.; Casson, L.K.; Pierce, W.M.; Qian, N.; Thomas, S.D.; Bates, P.J. As1411 alters the localization of a complex containing protein arginine methyltransferase 5 and nucleolin. Cancer Res. 2007, 67, 10491–10500.

31. Daniels, D.A.; Chen, H.; Hicke, B.J.; Swiderek, K.M.; Gold, L. A tenascin-c aptamer identified by tumor cell selex: Systematic evolution of ligands by exponential enrichment. Proc. Natl. Acad. Sci. USA 2003, 100, 15416–15421.

32. Borbas, K.E.; Ferreira, C.S.M.; Perkins, A.; Bruce, J.I.; Missailidis, S. Design and synthesis of mono- and multimeric targeted radiopharmaceuticals based on novel cyclen ligands coupled to anti-muc1 aptamers for the diagnostic imaging and targeted radiotherapy of cancer. Bioconjug. Chem. 2007, 18, 1205–1212.

33. Ferreira, C.S.M.; Matthews, C.S.; Missailidis, S. DNA aptamers that bind to muc1 tumour marker: Design and characterization of muc1-binding single-stranded DNA aptamers. Tumor Biol. 2006, 27, 289–301.

34. Wettstein, G.; Bellaye, P.S.; Micheau, O.; Bonniaud, P. Small heat shock proteins and the cytoskeleton: An essential interplay for cell integrity? Int. J. Biochem. Cell Biol. 2012, 44, 1680–1686.

35. Heng, Y.W.; Koh, C.G. Actin cytoskeleton dynamics and the cell division cycle. Int. J. Biochem. Cell Biol. 2010, 42, 1622–1633.

36. Hernandez, P.; Tirnauer, J.S. Tumor suppressor interactions with microtubules: Keeping cell polarity and cell division on track. Dis. Model. Mech. 2010, 3, 304–315.

37. Bruchez, M.; Moronne, M.; Gin, P.; Weiss, S.; Alivisatos, A.P. Semiconductor nanocrystals as fluorescent biological labels. Science 1998, 281, 2013–2016.

38. Ybo, J.; Kambara, T.; Gonda, K.; Higuchi, H. Intracellular imaging of targeted proteins labeled with quantum dots. Exp. Cell Res. 2008, 314, 3563–3569.

39. Xu, H.; Chen, C.A.; Peng, J.; Tang, H.W.; Liu, C.M.; He, Y.; Chen, Z.Z.; Li, Y.; Zhang, Z.L.; Pang, D.W. Evaluation of the bioconjugation efficiency of different quantum dots as probes for immunostaining tumor-marker proteins. Appl. Spectrosc. 2010, 64, 847–852.

40. Smith, B.R.; Cheng, Z.; De, A.; Koh, A.L.; Sinclair, R.; Gambhir, S.S. Real-time intravital imaging of rgd-quantum dot binding to luminal endothelium in mouse tumor neovasculature. Nano Lett. 2008, 8, 2599–2606.

41. Rhyner, M.N.; Smith, A.M.; Gao, X.H.; Mao, H.; Yang, L.L.; Nie, S.M. Quantum dots and multifunctional nanoparticles: New contrast agents for tumor imaging. Nanomedicine 2006, 1, 209–217.

42. Juzenas, P.; Chen, W.; Sun, Y.P.; Coelho, M.A.N.; Generalov, R.; Generalova, N.; Christensen, I.L. Quantum dots and nanoparticles for photodynamic and radiation therapies of cancer. Adv. Drug Deliv. Rev. 2008, 60, 1600–1614.

43. Bakalova, R.; Ohba, H.; Zhelev, Z.; Nagase, T.; Jose, R.; Ishikawa, M.; Baba, Y. Quantum dot anti-cd conjugates: Are

they potential photosensitizers or potentiators of classical photosensitizing agents in photodynamic therapy of cancer? Nano Lett. 2004, 4, 1567–1573.

44. Morosini, V.; Bastogne, T.; Frochot, C.; Schneider, R.; Francois, A.; Guillemin, F.; Barberi-Heyo, M. Quantum dot-folic acid conjugates as potential photosensitizers in photodynamic therapy of cancer. Photochem. Photobiol. Sci. 2011, 10, 842–851.

45. Samia, A.C.S.; Chen, X.B.; Burda, C. Semiconductor quantum dots for photodynamic therapy. J. Am. Chem. Soc. 2003, 125, 15736–15737.

46. Bagalkot, V.; Zhang, L.; Levy-Nissenbaum, E.; Jon, S.; Kantoff, P.W.; Langer, R.; Farokhzad, O.C. Quantum dot–aptamer conjugates for synchronous cancer imaging, therapy, and sensing of drug delivery based on bi-fluorescence resonance energy transfer. Nano Lett. 2007, 7, 3065–3070.

47. Gillies, E.R.; Frechet, J.M.J. Ph-responsive copolymer assemblies for controlled release of doxorubicin. Bioconjug. Chem. 2005, 16, 361–368.

48. Klostranec, J.M.; Chan, W.C.W. Quantum dots in biological and biomedical research: Recent progress and present challenges. Adv. Mater. 2006, 18, 1953–1964.]

49. Zhang, H.; Yee, D.; Wang, C. Quantum dots for cancer diagnosis and therapy: Biological and clinical perspectives. Nanomedicine 2008, 3, 83–91.

50. Liang, G.L.; Ren, H.J.; Rao, J.H. A biocompatible condensation reaction for controlled assembly of nanostructures in living cells. Nat. Chem. 2010, 2, 54–60.

11

CHAPTER

FABRICATION OF LOW-DENSITY GAN/ALN QUANTUM DOTS VIA GAN THERMAL DECOMPOSITION IN MOCVD

Jin Zhang[1], Senlin Li[1], Hui Xiong[1], Wu Tian[1], Yang Li[1], Yanyan Fang[1], Zhihao Wu[1], Jiangnan Dai[1*], Jintong Xu[2], Xiangyang Li[2] and Changqing Chen[1*]

(1)Wuhan National Laboratory for Optoelectronics, School of Optical and Electronic Information, Huazhong University of Science and Technology

(2)Key Laboratory of Infrared Imaging Materials and Detectors, Shanghai Institute of Technical Physics, Chinese Academy of Sciences

ABSTRACT

With an appropriate high anneal temperature under H_2 atmosphere, GaN quantum dots (QDs) have been fabricated *via* GaN thermal decomposition in metal organic chemical vapor deposition (MOCVD). Based on the characterization of atomic force microscopy (AFM), the obtained GaN QDs show good size distribution and have a low density of 2.4×10^8 cm^{-2}. X-ray photoelectron spectroscopy (XPS) analysis demonstrates that the GaN QDs were formed without Ga droplets by thermal decomposition of GaN.

Keywords: GaN quantum dots Thermal decomposition Low density MOCVD

BACKGROUND

Low-dimensional III-nitrides materials have gained much research attention because of their strong carrier confinement which may lead to the realization of next-generation electronic and optoelectronic applications [1–5]. Among these low-dimensional III-nitride materials, the study of single GaN quantum dot has become the recent focus due to its promising applications in the solid-state quantum computation, single-photon sources, and single-photon detectors, in which the density of quantum dots is required to be as low as approximately 10^8 cm^{-2}[6–9]. However, challenges remains in fabrication of low-density GaN quantum dots (QDs) with high quality. On the one hand, the most frequently used fabrication approach is self-assembly process *via* Stranski-Krastanov (SK) growth mode which requires sufficient lattice mismatch, but it is harder to acquire low-density GaN QDs and usually results in randomly distributed QDs with different sizes [10, 11]. On the other hand, although some low-density GaN nanodots can be obtained by the droplet epitaxy technique based on a vapor-liquid-solid process which offers distinct advantages in size and density manipulation of QDs, the droplet epitaxy technique usually results in QDs with the incomplete transition from Ga droplet to crystal GaN. What is more, there is almost no report about fabrication of low-density GaN QDs *via* the droplet epitaxy technique [12, 13]. Motivated by the above issues, recently, we have demonstrated the fabrication of GaN nanodots on AlN templates *via* GaN thermal decomposition in H_2 atmosphere, which does not involve the induction of strain or the crystallization of the Ga droplets [14]. In addition, the recent studies and applications of GaN-based materials growth have been demonstrated [15–20]. In this letter, the thermal decomposition conditions are further optimized and low-density GaN/AlN QDs with high quality are achieved. This study provides an alternative approach to fabricating low-density GaN QDs for single-photon devices.

METHODS

GaN QDs were formed on AlN/sapphire templates by metal organic chemical vapor deposition (MOCVD). Triethylgallium (TEGa), trimethylaluminum (TMAl), and ammonia were used as precursors for Ga, Al, and N sources with H_2 as carrier gas. The total pressure was maintained at 40 Torr. The sapphire substrates were introduced into the MOCVD reactor and 800-nm-thick AlN buffer layers were deposited. Then, 800-nm-thick GaN epilayers were grown on the AlN templates at 940°C. Subsequently, the samples were annealed in H_2 atmosphere at different conditions: sample A, at 1,050°C for 5 min; sample B, at 1,100°C for 5 min and sample C, at 1,100°C for 8 min, as shown in Figure 1. After that, these samples were cooled down to room temperature at the presence of $NH_3 + H_2$. Besides, two controlled experiments were also conducted. One was the growth of 800-nm-thick GaN on 800-nm-thick AlN/sapphire without decomposition in H_2 (sample D), and another one is an 800-nm-thick AlN buffer template on sapphire without decomposition in H_2 (sample E). The surface morphologies of all samples were characterized by atomic force microscopy (AFM) measurements. The surface chemistries of obtained GaN QDs and some control samples were investigated using X-ray photoelectron spectroscopy (XPS) measurements with monochromatic Mg Kα X-ray source (hν = 1,253.6 eV).

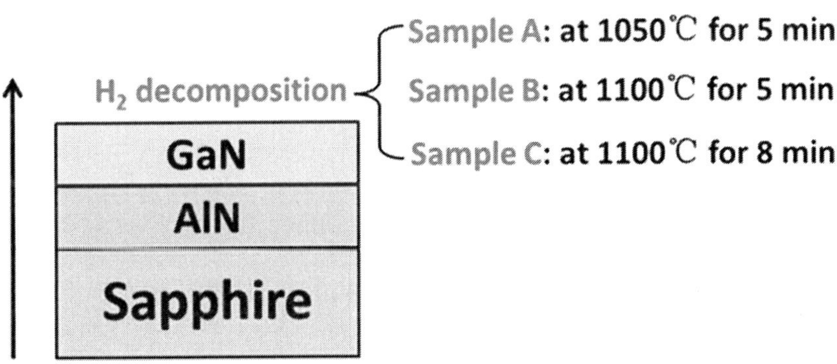

Figure 1. The schematic of h 2 -annealed conditions of samples a, b, and c.

RESULTS AND DISCUSSION

The surface morphologies of all samples were studied by atomic force microscopy (AFM), and the results are shown in Figure 2. Compared with the surface morphology of controlled sample D (Figure 2d), it is obvious that GaN decomposition occurs for Sample A (Figure 1a). Figure 1f is the corresponding three-dimensional (3D) AFM image of Figure 1a, in which distributed dots are on terraces and abrupt peaks are to be buried in the side wall, indicating the decomposition process for the formation of GaN dots. As the decomposition occurred toward the inner of the side wall, the abrupt peaks are then exposed to H_2 flow and decomposed. Since the heights of peaks decrease faster than the diameters of peaks, the side wall is etched away and the peaks are etched to small dots with a longer etching time, which is consistent with our previous observation [14]. With increasing of the annealing temperature from 1,050°C to 1,100°C, the decomposition of GaN has an interesting phenomenon that the steps disappear and well-shaped dots are just left on a flat surface, as shown in Figure 2b. The obtained GaN QDs show a low density in the magnitude of approximately 10^8 cm^{-2}. As expected, these dots are etched as the elongation of annealing time from 5 to 8 min, left with atomically flat surface (Figure 2c) similar to that of controlled sample E (Figure 2e). It is clear that surface morphology of the AlN buffer templates before and after annealing in H_2 are exactly the same, indicating that no decomposion of AlN takes place at the temperature of 1,100°C. This result is in good agreement with the claim made by Y. Kumagai et al. [21].

To further investigate the size distribution of the obtained GaN QDs, the AFM images of sample B with scan area 10×10 μm^2 is shown in Figure 3a. The QDs have a low density of approximately 2.4×10^8 cm^{-2} and no obvious big dots are observed, showing the good uniformity. Figure 3b illustrates the diameter histograms obtained by the AFM analysis of Figure 3a. It is clear that the most probable diameter is in the range from 70 to 80 nm. The inset of Figure 3b shows a detailed 3D AFM image of the QDs in 1×1 μm^2, indicating the similar well-formed dot structure. According to the results above, the obtained GaN QDs have a good size distribution. To the best of our knowledge, this is the first report of low-density GaN QDs fabricated *via* GaN thermal decomposition in MOCVD.

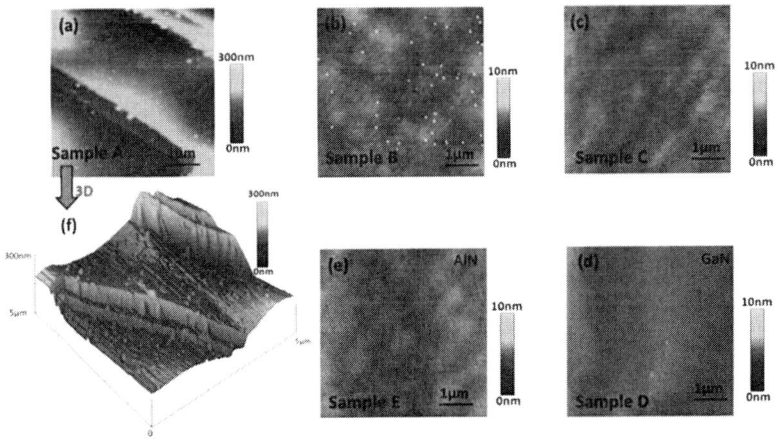

Figure 2. Afm images of samples. Samples **(A)** A, **(B)** B, **(C)** C, **(D)** D, **(E)** E, and **(F)** corresponding the 3D AFM image of sample A.

As is shown in Figure 4, since XPS analysis was performed for samples A, B, and C, Ga2*p* and N1*s* core level spectra were measured. For both of the XPS spectra, the C1*s* peak at approximately 285.0 eV was used as binding-energy reference. Baselines were fixed using a Shirley background subtraction model and all peaks were fitted using a linear combination of 80% Gaussian and 20% Lorentzian line shapes. On the one hand, the Ga2*p* spectra are analyzed in Figure 4a. Both samples A and B have a Ga2*p* peak which can be fitted as only one subpeak located at 1,117.1 eV, which is assigned to Ga-N bond [22–24]. So there are no Ga droplets but GaN on the surface of samples A and B, indicating that the Ga desorption rate exceed the GaN decomposition rate. On the contrary, if the Ga desorption rate is less than the GaN decomposition rate, Ga droplets will generate in a chemical manner and Ga-Ga bond will be observed. No Ga2*p* peaks were observed in sample C, confirming that sample C is just the AlN

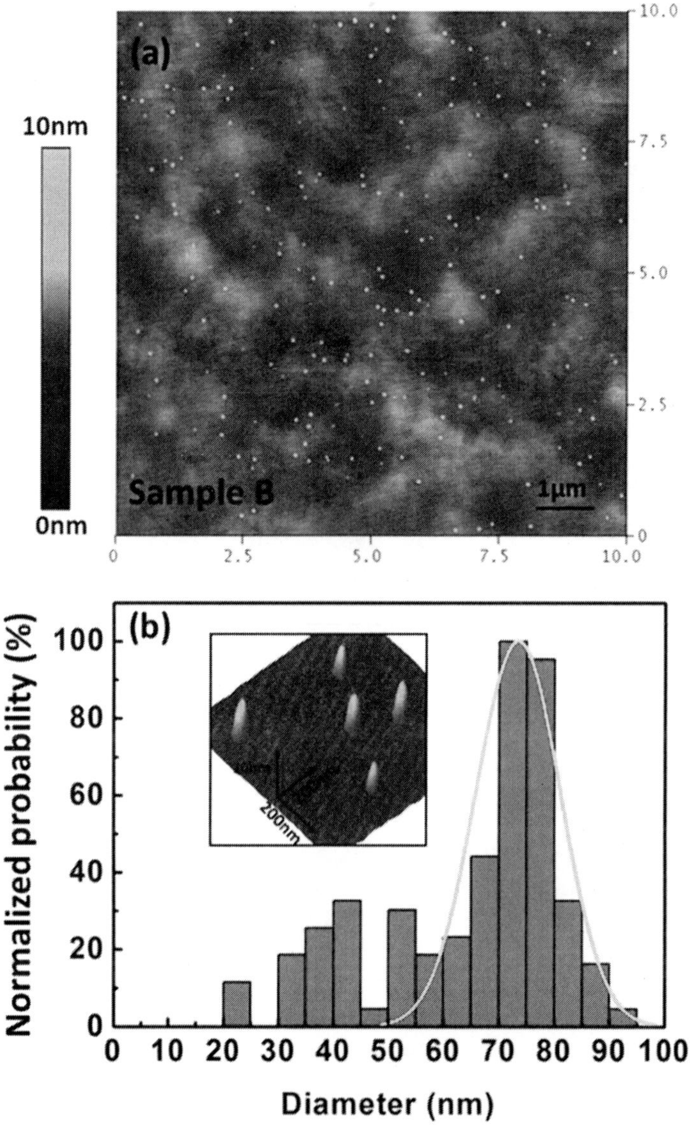

Figure 3. AFM IMAGES OF SAMPLE B (A) AND DIAMETER DISTRIBUTIONS OF GAN QDS **(B). (A)** Scan area 10 × 10 μm²; **(B)** Analyzed from the AFM images of sample B. Inset is the 3D image of obtained GaN QDs.

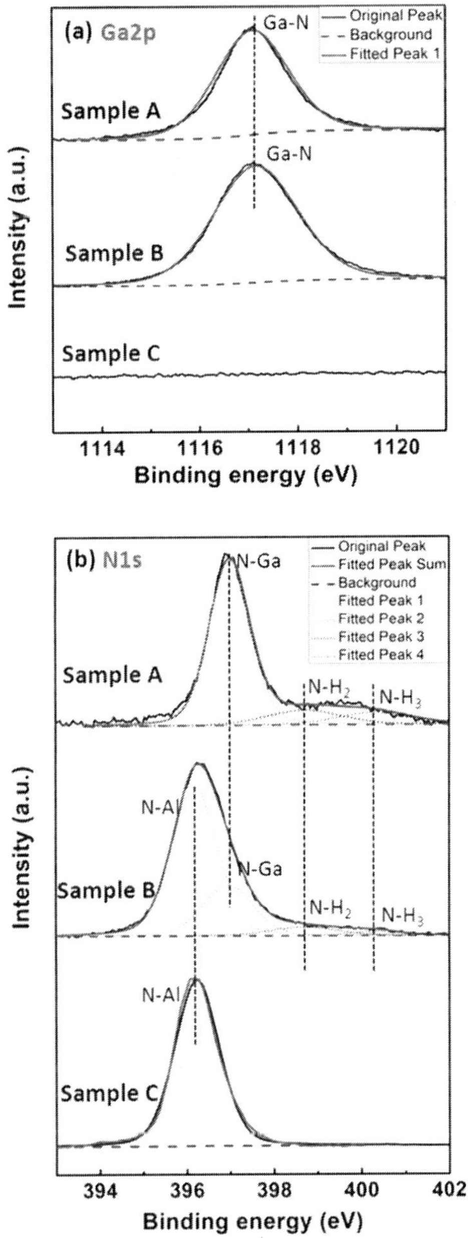

Figure 4. XPS SPECTRA OF (A) GA2 **p** AND (B) N1 **s** FOR SAMPLES A, B, AND C. The background lines and the fitted lines were also subtracted.

buffer after H_2 decomposition. On the other hand, the N1*s* spectra are analyzed in Figure 4b. For sample A, the N1*s* spectra can be decomposed into a total of four fitted subpeaks at 397.0, 398.7, and 400.3 eV, which were assigned to N-Ga bond, N-H_2 bond and N-H_3 bond [25, 26], respectively. Only GaN existed on the surface of sample A. For sample C, the N1*s* spectra can be decomposed into one subpeaks at 398.7 eV, which is assigned to N-Al bond [27]. Only AlN existed on the surface of sample C. For sample B, the N1*s* spectra were decomposed into a total of four fitted subpeaks at 396.2, 397.0, 398.7, and 400.3 eV, which can be assigned to N-Al bond, N-Ga bond, N-H_2 bond, and N-H_3 bond, respectively. These fitted subpeaks coincide with the fitted subpeaks of samples A and C, providing a chemical evidence for the existence of GaN QDs formed on the AlN buffer. In addition, the N-H_2 bond and N-H_3 bond were obtained in samples A and B but did not exist in sample C, indicating that the appearance of N-H_2 bond and N-H_3 bond were caused by the interaction of decomposed GaN and hydrogen at high temperature.

CONCLUSIONS

In summary, GaN QDs were fabricated on flat AlN buffer *via* H_2 decomposition of GaN film at high temperature. The AFM measurements show that the obtained GaN QDs have good size uniformity and a low dot density about 2.4×10^8 cm^{-2}. The XPS spectra analysis actually demonstrated that the GaN QDs do not contain Ga droplets. The results provide an alternative approach to fabricate low-density GaN QDs for applications in single-photon devices.

AUTHORS' INFORMATION

JZ, SLL, WT, and YL are PhD students, HX is the postdoctor, JND, YYF and ZHW hold associate professor positions, and CQC is a professor at the Huazhong University of Science and Technology. XYL and JTX hold the researcher and associate researcher positions at the Shanghai Institute of Technical Physics.

ABBREVIATIONS

AFM: atomic force microscopy

3D: three-dimensional

MOCVD: metal-organic chemical vapor deposition

QDs: quantum dots

S-K: Stranski-Krastanov

TEGa: triethylgallium

TMAl: trimethylaluminum

XPS: X-ray photoelectron spectroscopy.

ACKNOWLEDGEMENTS

This work was supported by the National Basic Research Program of China (Grant Nos. 2012CB619302 and 2010CB923204) and in part by the foundation of the Science and Technology Bureau of Wuhan City (Grant No. 2014010101010006).

AUTHORS' CONTRIBUTIONS

JZ wrote the manuscript and participated in all the experiments and the data analysis. SLL, HX, WT, YL, ZHW, and JND partially participated in the experiments and the data analysis. JTX and XYL offer supporting in the testing of XPS. YYF and CQC supervised the writing of the manuscript and all the experiments. All authors read and approved the final manuscript.

REFERENCES

1. Kawasaki K, Yamazaki D, Kinoshita A, Hirayama H, Tsutsui K, Aoyagi Y: GaN quantum-dot formation by self-assembling droplet epitaxy and application to single-electron transistors. Appl Phys Lett 2000, 79: 2243–2245.
2. Schupp T, Meisch T, Neuschl B, Feneberg M, Thonke K, Lischka K, As DJ: Droplet epitaxy of zinc-blende GaN quantum dots. J Crystal Growth 2010, 312: 3235–3237. 10.1016/j.jcrysgro.2010.07.049
3. Li S, Ware M, Wu J, Minor P, Wang Z, Wu Z, Jiang Y, Salamo GJ: Polarization induced pn-junction without dopant in graded AlGaN coherently strained on GaN. Appl Phys Lett 2012, 101: 122103. 10.1063/1.4753993
4. Li S, Zhang T, Wu J, Yang Y, Wang Z, Wu Z, Chen Z, Jiang Y: Polarization induced hole doping in graded AlxGa1-xN (x = 0.71) layer grown by molecular beam epitaxy. Appl Phys Lett 2013, 102: 062108. 10.1063/1.4792685
5. Li S, Ware ME, Wu J, Kunets VP, Hawkridge M, Minor P, Wang Z, Wu Z, Jiang Y, Salamo GJ: Polarization doping: reservoir effects of the substrate in AlGaN graded layers. J Appl Phys 2012, 112: 053711. 10.1063/1.4750039
6. Michler P: Single Semiconductor Quantum Dots. Heidelberg: Springer; 2009.
7. DiVincenzo DP: Double quantum dot as a quantum bit. Science 2005, 309: 2173–2174. 10.1126/science.1118921
8. Mowbray DJ, Skolnick MS: New physics and devices based on self-assembled semiconductor quantum dots. J Phys D Appl Phys 2005, 38: 2059–2076. 10.1088/0022-3727/38/13/002
9. Liang CT, Simmons MY, Smith CG, Kim G-H, Ritchie DA, Pepper M: Multilayered gated lateral quantum dot devices. Appl Phys Lett 2000, 76: 1134–1136. 10.1063/1.125961
10. Shchukin VA, Ledentsov NN, Bimberg D: Epitaxy of Nanostructures. N.Y.: Springer Verlag; 2003.

11. Lee J, Wang ZM, Hirono Y, Dorogan VG, Mazur YI, Kim ES, Koo SM, Park S, Song S, Salamo GJ: Low-density quantum dot molecules by selective etching using in droplet as a mask. IEEE Trans Nanotechnol 2011, 10: 600–605.

12. Maruyama T, Otsubo H, Kondo T, Yamamoto Y, Naritsuk S: Fabrication of GaN dot structure by droplet epitaxy using NH_3. J Cryst Growth 2007, 301: 486–489.

13. Debnath RK, Stoica-a T, Besmehn A, Jeganathan K, Sutter E, Meijers R, Luth H, Calarco R: Formation of GaN nanodots on Si (111) by droplet nitridation. J Cryst Growth 2009, 311: 3389–3394. 10.1016/j.jcrysgro.2009.04.025

14. Xiong H, Zhang J, Li SL, Wang H, Fang YY, Dai JN, Chen CQ: Fabrication of GaN nanodots via GaN thermal decomposition in H_2 atmosphere. J Vac Sci Technol B 2013, 31: 050603–050607. 10.1116/1.4817499

15. Chen YS, Liao CH, Chueh YL, Kuo CT, Wang HC: Plan-view transmission electron microscopy study on coalescence overgrowth of GaN nano-columns by MOCVD. Opt Mat Express 2013, 3: 1459–1467. 10.1364/OME.3.001459

16. Chen YS, Liao CH, Cheng YC, Kuo CT, Wang HC: Nanostructure study of the coalescence growth of GaN columns with molecular beam epitaxy. Opt Mat Express 2013, 3: 1450–1458. 10.1364/OME.3.001450

17. Feng SW, Tu LW, Wang HC, Sun Q, Han J: The role of growth-pressure on the determination of anisotropy properties in nonpolar m-plane GaN. ECS J Solid State Sci Technol 2012, 1: R50-R53.

18. Feng SW, Lin HC, Chyi JI, Tsai CY, Huang CJ, Wang HC, Yang FW, Lin YS: The impact of trimethylindium treatment time during growth interruption on the carrier dynamics of InGaN/GaN multiple quantum wells. Thin Solid Films 2011, 519: 6092–6096. 10.1016/j.tsf.2011.04.004

19. Wang HC, Tang TY, Yang CC, Malinauskas T, Jarasiunas K: Carrier dynamics in coalescence overgrowth of GaN

nanocolumns. Thin Solid Films 2010, 519: 863. 10.1016/j.tsf.2010.08.149

20. Wang HC, Malinauskas T, Jarasiunas K, Feng SW, Ting CC, Liu S: Carrier dynamics in InGaN/GaN multiple quantum wells based on different polishing processes of sapphire substrate. Thin Solid Films 2010, 518: 7291. 10.1016/j.tsf.2010.04.093

21. Kumagai Y, Akiyama K, Togashi R, Murakami H, Takeuchi M, Kinoshita T, Takada K, Aoyagi Y, Koukitu A: Polarity dependence of AlN {0 0 0 1} decomposition in flowing H_2. J Crys Growth 2007, 305: 366–371. 10.1016/j.jcrysgro.2007.04.005

22. Choi HW, Cheong MG, Rana MA, Chua SJ, Osipowicz T, Pan SJ: Rutherford backscattering analysis of GaN decomposition. J Vac Sci Technol B 2003, 21: 1080–1083. 10.1116/1.1577570

23. Choi HW, Rana MA, Chua SJ, Sipowicz TO, Pan JS: Surface analysis of GaN decomposition. Semicond. Sci Technol 2002, 17: 1223–1225. 10.1088/0268-1242/17/12/304

24. Kuriyama K, Tsunoda T, Hayashi N, Yukimi T: Characterization of GaN synthesized in N-ion implanted GaAs. Phys Res B 1999, 148: 432–436.

25. Carin R, Deville JP, Werckmann J: An XPS study of GaN thin films on GaAs. Surf Interface Anal 1990, 16: 65–69. 10.1002/sia.740160116

26. Li D, Sumiya M, Fuke S, Yang D, Que D, Suzuki Y, Fukuda Y: Selective etching of GaN polar surface in potassium hydroxide solution studied by X-ray photoelectron spectroscopy. J Appl Phys 2001, 90: 4219–4223. 10.1063/1.1402966

27. Rosenberger L, Baird R, McCullen E, Auner G, Shrevea G: XPS analysis of aluminumnitride films deposited by plasma source molecular beam epitaxy. Surf Interface Anal 2008, 40: 1254–1261. 10.1002/sia.2874

12

CHAPTER

ASSEMBLY OF CDS QUANTUM DOTS ONTO HIERARCHICAL TIO2 STRUCTURE FOR QUANTUM DOTS SENSITIZED SOLAR CELL APPLICATIONS

Syed Mansoor Ali [1,*], Mohamed Aslam [1], W. A. Farooq [1], Amanullah Fatehmulla [1] and M. Atif [1,2,*]

[1] *Department of Physics and Astronomy, College of Sciences, P.O. Box 2455, King Saud University, Riyadh 11451, Saudi Arabia*

[2] *National Institute of Laser and Optronics, Nilore, Islamabad 45650, Pakistan*

ABSTRACT

Quantum dot (QD) sensitized solar cells based on Hierarchical TiO2 structure (HTS) consisting of spherical nano-urchins on transparent conductive fluorine doped tin oxide glass substrate is fabricated. The hierarchical TiO2 structure consisting of spherical nano-urchins on transparent conductive fluorine doped tin oxide glass substrate synthesized by hydrothermal route. The CdS quantum dots were grown by the successive ionic layer adsorption and reaction deposition method. The quantum dot sensitized solar cell based on the hierarchical TiO2 structure shows a current density JSC = 1.44

mA, VOC = 0.46 V, FF = 0.42 and η = 0.27%. The QD provide a high surface area and nano-urchins offer a highway for fast charge collection and multiple scattering centers within the photoelectrode.

Keywords: hierarchical TiO2 structure; nanourchins; CdS quantum dots; FETEM; FESEM; UV-Vis spectroscopy

1. INTRODUCTION

In the modern age, solar cells have attracted significant attention due to their promising applications in energy generation devices. Since the pioneering report by O'Regan and Grätzel in 1991, dye-sensitized solar cells have been investigated extensively all over the world [1,2,3,4,5,6,7,8,9,10,11]. The quantum dot sensitized solar cell (QDSSC) has received wide attentions recently because they have several advantages over dye sensitizers, such as tunable energy gaps [12], high absorption coefficients [13], and generation of multiple electron-hole pair with high energy excitation [14].

The TiO2 nanoparticle based photoelectrode showed considerable power conversion efficiency over a large surface area with the attachment of dye molecules. M. Pavan et al. reported on an oxide heterojunction solar cell, entirely produced by spray pyrolysis onto fluorine doped tin oxide (FTO) covered glass substrates [15]. However, the irregular stacking of TiO2 nanocrystallines have been found to limit the electron transportation and decreases the electron life time because of the random network of crystallographically misaligned crystallites, and lattice mismatches at the grain boundaries [15,16,17,18]. It has been accepted that the value of power conversion efficiency of photoelectrodes highly depends on the morphology and structure of TiO2. In order to increase the photovoltaic performance, through their excellent electron transport and light scattering ability, one-dimensional nanostructures, such as nanorods (NRs), nanowires (NWs) or nanotubes (NTs), have been studied as photoelectrode materials for sensitized solar cells [19,20,21,22,23]. Due to low specific surface area ascribing to larger diameter and wide gaps among neighbor NWs [23,24], the TiO2 NWs based photoelectrode has not shown remarkable enhancement of power conversion efficiency. To overcome this problem, hierarchically-structured materials composed of nanocrystallites that form large micro-spheres. Nanocrystallites can provide excellent light scattering with large surface area for sensitizer-uptake. In these hierarchical materials, slow trap-limited charge transport remains a fundamental problem. To solve this problem, a nanourchin (NU) TiO2 are formed by clustering nanowires that have a mean diameter of about 50 nm and a length of a few micrometers to construct a radially aligned structure. There are few recent examples concerning hierarchical TiO2 structure (HTS), such as either for rutile TiO2 on FTO glass or anatase TiO2 on a Ti foil substrate for improving the power conversion efficiency [19,24].

In the present work, hierarchical TiO2 structure (HTS) consisting of spherical nano-urchins was synthesized through hydrothermal method. The CdS QDs were assembled by successive ion layer adsorption and reaction (SILAR). The HTS/CdS QDs based photoelectrode was used to improve the power conversion efficiency of quantum dot sensitized solar cell.

2. EXPERIMENTAL

2.1. Synthesis of Hierarchical TiO2 Structure

The hierarchical TiO2 structure (HTS) was grown on the FTO substrate. In a typical synthesis, the substrate was ultrasonically cleaned sequentially in acetone, isopropyl alcohol and deionized water for 15 min and was finally dried with nitrogen flow. Separately, 1 mL of titanium isopropoxide was added dropwise to a 1:1 mixture of deionized water and concentrated (35%) hydrochloric acid to obtain a clear transparent solution. The substrate was placed at an angle in a 100 mL Teflon liner, and the precursor solution was added to it. The Teflon liner was loaded in an autoclave and was placed in furnace. The growth was carried out at 180 °C for 15 h.

2.2. Deposition of CdS Quantum Dots (QD)

The CdS quantum dots were deposited on HTS films by successive ionic layer deposition and reaction (SILAR) method. The HTS electrode was exposed to Cd2+ and S2− ion successive immersion in a ethanolic solutions of 0.5 M Cd(NO3)2 and methnolic solution of 0.5 M Na2S. The film was dipped into 0.5 M Cd(NO3)2 solution for 1 min and rinsed with ethanol and then, dipped into 0.5 M Na2S for 1 min and rinsed with methanol. These dipping procedures are considered one cycle. The coating procedure was repeated 10 times. For the deposition of CdS on the HTS by SILAR method, the experimental procedure is explained as follow:

$$\text{(HTS)} \xrightarrow{Cd^{2+}} \text{(HTS)}Cd^{2+} \longrightarrow \text{Rinsed} \xrightarrow{S^{2-}} \text{(HTS)CdS} \longrightarrow \text{Rinsed}$$

2.3. Preparation of Electrolyte Solution

Polysulfide electrolytes were prepared by mixing suitable quantities of Na2S, S, and KCl powders in water/methanol solution taken in the ratio 3/7.

2.4. Fabrication of QDSCs

The QD-adsorbed HTS was used as the working electrode and platinum coated FTO glass as counter electrode. The electrodes were assembled into a sealed cell with a cello tape spacer and binder clips, with an active area equal to 0.36 cm2. The electrolyte was injected from the edges into the open cell, and the cell was tested. The schematic of the studied QDSSCs is shown in Figure 1.

2.5. Characterization

X-Ray diffraction XRD analysis of HTS/FTO films was carried out using multipurpose X-ray diffractometer (Bruker D8 Discover, Bruker AXS GmbH, Karlsruhe, Germany) with Cukα source radiation. Surface morphology of the films was investigated with a JEOL (JSM-7600F) Field Emission electron microscope (Jeol, Peabody, MA, USA). Size of the CdS QDs was measured by JEOL (JEM-2100F) field emission electron microscope (FETEM) (Jeol, Peabody, MA, USA).

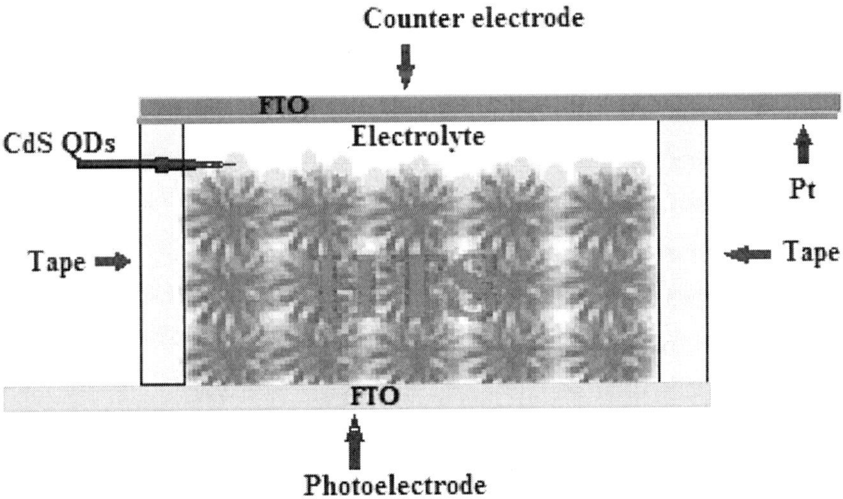

Figure 1. Schematic of the Hierarchical TiO2 structure (HTS) based quantum dot sensitized solar cell (QDSSC).

Optical absorption studies were made at room temperature by using UV-Vis-NIR spectrophotometer (JASCO-V 670) (Jasco, Halifax, NS, Canada) in the wavelength range 200 nm–800 nm. The current-voltage and capacitance-voltage characteristics were investigated using a Semiconductor Characterization System SC-4200 from Keithley (Keithley Instruments, Solon, OH, USA). The films were illuminated by a Class-BBA Solar Simulator (PV measurements, Boulder, CO, USA) and TM-206 solar power meter (Tenmars, Taipei, Taiwan) was used for measuring the light intensity.

3. RESULTS AND DISCUSSION

Figure 2 shows the XRD patterns of TiO2 nanowires grown on FTO glass substrate. Nanowires grown in this work, regardless of substrate used, were found to have the rutile phase by matching between the observed and standard "d" value of the TiO2 nanostructure. XRD data (Figure 1) show a good agreement with the standard TiO2 (PDF file #01-086-0147, P42/mnm, a = b = 4.594 Å and c = 2.958 Å). In XRD spectrum, the respective diffraction peaks corresponding to the FTO are denoted by the symbol "F".

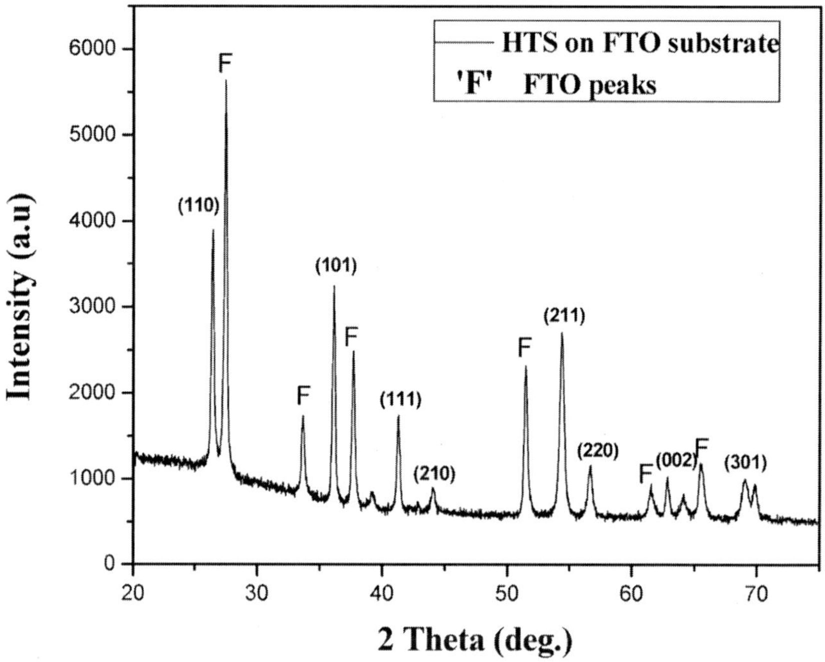

Figure 2. X-ray diffraction pattern of hierarchical TiO2 structure on fluorine doped tin oxide (FTO) glass substrate.

The field emission scanning electron microscope (FESEM) images of TiO2 nanostructure on FTO glass is shown in Figure 3. From Figure 3a,b we have found that the morphology of TiO2 is a hierarchical structure. It is believed that a hierarchical TiO2 structure (HTS) has three novel levels. The first level, the TiO2, is made up of nano-urchins (NUs); the second level, the NUs, is composed of nanowires (NWs); and the third level, the NWs, is made up of nanoparticles (NPs) [25]. Figure 3a depicts the NUs grown on the FTO glass substrate with an average diameter of 8 μm. Higher magnification of FESEM image Figure 3b reveals the TiO2 nano-urchins is composed of the TiO2 nanowires of average length of 1 μm.

Figure 3. Field emission scanning electron microscopy images of hierarchical TiO2 structure. (a) at low resolution (1000 X); (b) at high resolution (10,000 X).

Figure 4a,b shows the low and high magnification TEM images of CdS QDs nanoparticles deposited by SILAR technique on the TiO2 nanostructure. A clear morphology of CdS QDs nanoparticles, with a size from 4 nm to 6 nm, indicates that CdS QDs are markedly immersed on the surface of TiO2 nanostructure.

The comparison of the absorption spectra of HTS and CdS QDs deposited on the HTS is shown in Figure 5. The immersion of CdS QDs in HTS structure has improved the optical absorbance in the visible region. The absorption edge obtained from the intersection of the sharply decreasing region of a spectrum with its baseline for HTS around 370 nm and shifted to longer wavelength around 520 nm after immersion of CdS QD. Corresponding to this absorption peak, the band gap was calculated to be 2.7 eV. The value reported for CdS in bulk was 2.42 eV [26]. The band gap of CdS particles deposited on HTS films was higher than that of CdS bulk, which indicated that the size of the CdS particles were still within the scale of quantum dot. Estimated from the absorption edge of the absorption spectra, the radius of CdS particles was calculated to be 2.37 nm by using the hyperbolic band model (HBM) equation [27].

$$R = \sqrt{\frac{h^2 E_{bulk}}{2m^*(E_{nano}^2 - E_{bulk}^2)}} \tag{1}$$

where Ebulk is bulk band gap; Enano is band gap of nanomaterial; m* is effective mass of electron in bulk CdS (m* = 0.21mo). Hence the particle size estimated as 2R is 4.74 nm.

A new approach implemented the benefits of one-dimensional nanostructures by suitably combining the NU TiO2 and nanoparticles (NP) to construct a hierarchical TiO2 structure (HTS) photoelectrode for the QDSSCs. In particular, the QD provide a high surface area for sufficient QD deposition by SILAR technique, whereas NU TiO2 particles offer a highway for fast charge collection and multiple scattering centers within the photoelectrode. The QDSSCs made of the HTS film exhibited remarkable improvement in power conversion efficiency in comparison to the reference cell made with the NP film [28,29,30].

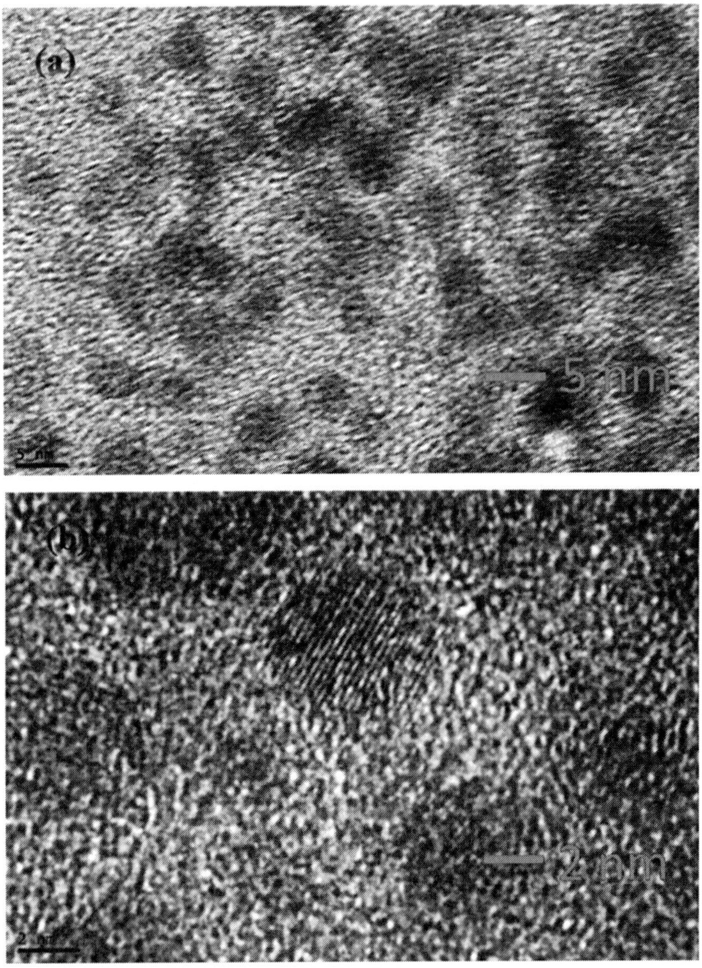

Figure 4. Field emission transmission electron microscopy images of CdS QDs in (a) low resolution and 5 nm (b) High resolution 2 nm.

To evaluate the photovoltaic performance of the HTS, the synthesized products were applied as a photoanode for QDSSC application. The J–V characteristics of the quantum dot sensitized (FTO/HTS/CdS QDs/Pt/FTO) solar cell were measured using a solar simulator. The fourth quadrant of J–V and P–V characteristics are shown in Figure 6. The photocurrent is defined as a current produced under light irradiation due to the generation of free charge carriers by absorption of photons within the depletion layer. Figure 6 demonstrate that the value of J increases with the increasing illumination intensities and the photovoltage value increased up to 0.478 V at 90 mW/cm2 then decreases to 0.46 V at100 mW/cm2. The values of short circuit current density Jsc and open circuit voltage Voc are found to be 1.44 mA/cm2 and 0.46 V under 100 mW/cm2 illuminations, respectively, with an efficiency of 0.27%. As per the literature, we expected the assembly of CdS quantum dots onto hierarchical TiO2 structure for quantum dots sensitized solar cell to give good results. However, with our fabricated cell, we do not observe similar results. The probable reason for obtaining lower efficiency may be due to the QDs embedded in the TiO2 nano urchins matrix provides many pathways, resulting in an increase in injection time.

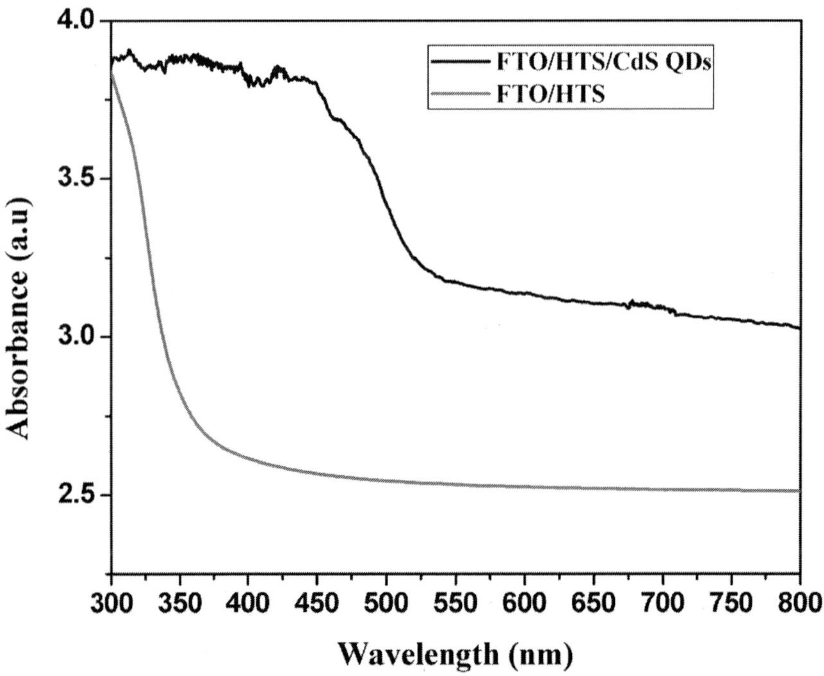

Figure 5. Absorbance spectra of HTS and HTS/CdS QDs.

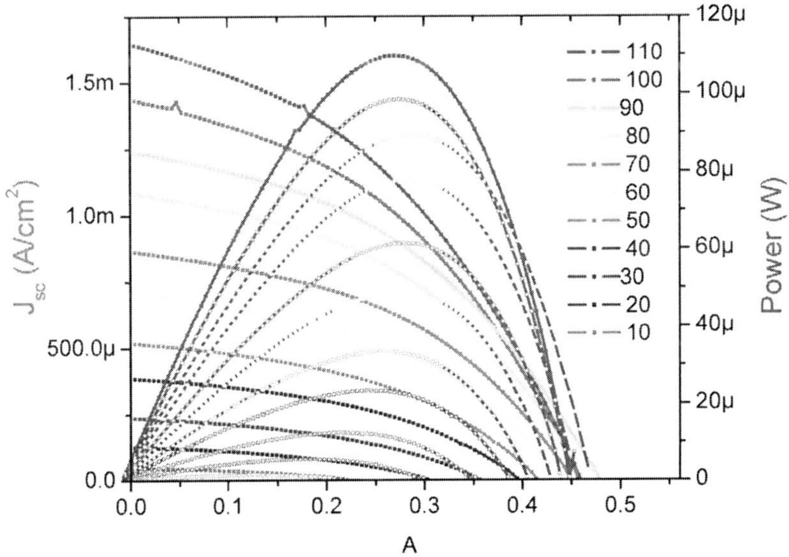

Figure 6. J–V and P–V characterization of HTS based QDSSC at different intensities of light.

By power-voltage curve, we can take more information about the delivered power to this device. The electric power first increases with the increasing of the voltage and reaches its maximum value and then, decreases and reaches to zero with the further increase of the voltage. The maximum value of power indicates how much the QDSSC can deliver as a maximum power to an external load and is defined as Pmax = IM × VM, where IM is the maximum current and VM is the maximum voltage at each value of illumination intensities.

Figure 7a shows the graph between the short circuit current density Jsc with open circuit voltage Voc. It is seen that the value of Voc increases exponentially with the increase of Isc. This exponentially increasing trend of Voc with Isc obeys the following relation [31,32].

$$V_{oc} = \frac{nkT}{q} ln\left(\frac{J_{sc}}{J_o} + 1\right) \tag{2}$$

where n is the diode ideality factor; k is the Boltzmann's constant; q is the electric charge; and Jo is the reverse saturation current density. By using the above equation, fitted to the graph of Voc–Jsc, the ideality factor of fabricated cell is found to be 3 [33]. For an ideal p-n junction, the ideality factor is considered to be approximately 1 under room temperature. The probable reasons for larger ideality factors may be due to several defects as well as local non-linear shunts anywhere in the cell area that are responsible the ideality factor value greater than 1 [34].

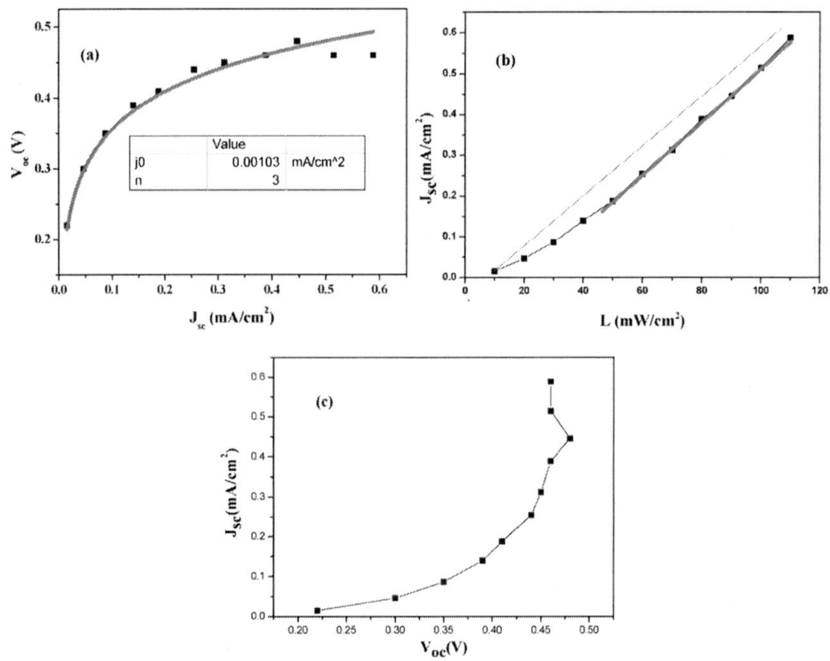

Figure 7. Photovoltaic performance parameter of HTS based QDSSC (a) Voc–Jsc; (b) Jsc–L; (c) Jsc–Voc plot.

The value of short circuit current density Jsc of the solar cell were obtained from J–V curves under light illumination. In Figure 7b it is observed that there exists a non-linear relation between current and light intensity in the low intensity region. This is due to the large shunt resistance in the device. The linear relation is fully recovered after the light intensity is increased to 40 mW/cm2, indicating that the photo filling effect has saturated the non-radiative recombination center.

Figure 7c show the Jsc–Voc relation plot under different light intensity and reveals that the dark diode property of the fabricated with high series resistance.

4. CONCLUSIONS

A novel hierarchical TiO2 structure on transparent conductive FTO glass substrate is synthesized by hydrothermal route. The CdS quantum dots were grown by the successive ionic layer adsorption and reaction deposition method. The investigations of FESEM reveal that HTS consist of spherical nano-urchins and FETEM indicate that CdS QDs are markedly immersed on the surface of TiO2 nanostructure. The QD provide a high surface area and nano-urchins offer a highway for fast charge collection and multiple scattering centers within the photoelectrode, which are responsible for the improvement of power conversion efficiency. We anticipate that nano-urchins consisting of HTS based photoelectrodes could be promising for the fabrication of high efficiency Perovskite quantum dot sensitized solar cell.

ACKNOWLEDGMENTS

This project was funded by the National Plan for Science, Technology and Innovation (MAARIFAH), King Abdulaziz City for Science and Technology, Kingdom of Saudi Arabia, Award Number 11NAN 1464-02.

AUTHOR CONTRIBUTIONS

All the authors contributed significantly for the completions of this manuscript. Mohamed Aslam and Syed Mansoor Ali fabricated solar cell and M. Atif, Amanullah Fatehmulla, W. A. Farooq participated in the experimental as well as data analysis. In short, the whole team contributed in writing up the manuscript.

REFERENCES

1. O'Regan, B.; Grätzel, M. A low-cost, high-efficiency solar cell based on dye-sensitized colloidal TiO2 films. Nature 1991, 335, 737–740.
2. Yin, X.; Xue, Z.S.; Liu, B. Electrophoretic deposition of Pt nanoparticles on plastic substrates as counter electrode for flexible dye-sensitized solar cells. J. Power Sources 2011, 196, 2422–2426.
3. Song, H.K.; Yoon, J.S.; Won, J.; Kim, H.; Yeom, M.S. New approach to the reduction of recombination in dye-sensitised solar cells via complexation of oxidised species. J. Nanosci. Nanotechnol. 2013, 13, 5136–5141.
4. Guo, W.X.; Xu, C.; Wang, X.; Wang, S.H.; Pan, C.F.; Lin, C.J.; Wang, Z.L. Rectangular bunched rutile TiO2 nanorod arrays grown on carbon fiber for dye-sensitized solar cells. J. Am. Chem. Soc. 2012, 134, 4437–4441.
5. Sun, X.M.; Sun, Q.; Li, Y.; Sui, L.N.; Dong, L.F. Effects of calcination treatment on the morphology and crystallinity, and photoelectric properties of all-solid-state dye-sensitized solar cells assembled by TiO2 nanorod arrays. Phys. Chem. Chem. Phys. 2013, 15, 18716–18720.
6. Kong, J.; Zhou, Z.J.; Li, M.; Zhou, W.H.; Yuan, S.J.; Yao, R.Y.; Zhao, Y.; Wu, S.X. Wurtzite copper-zinc-tin sulfide as a superior counter electrode material for dye-sensitized solar cells. Nanoscale Res. Lett. 2013, 8.
7. Grätzel, M. Solar energy conversion by dye-sensitized photovoltaic cells. Inorg. Chem. 2005, 44, 6841–6851.
8. Fan, X.; Chu, Z.Z.; Wang, F.Z.; Zhang, C.; Chen, L.; Chen, Y.; Tang, Y.W.; Zou, D.C. Wire-shaped flexible dye-sensitized solar cells. Adv. Mater. 2008, 20, 592–595.

9. Lee, M.R.; Eckert, R.D.; Forberich, K.; Dennler, G.; Brabec, C.J.; Gaudiana, R.A. Solar power wires based on organic photovoltaic materials. Science 2009, 324, 232–235.

10. Tsai, J.K.; Hsu, W.D.; Wu, T.C.; Meen, T.H.; Chong, W.J. Effect of compressed TiO2 nanoparticle thin film thickness on the performance of dye-sensitized solar cells. Nanoscale Res. Lett. 2013, 8.

11. Wang, D.; Hou, S.C.; Wu, H.W.; Zhang, C.; Chu, Z.Z.; Zou, D.C. Fiber-shaped all-solid state dye sensitized solar cell with remarkably enhanced performance via substrate surface engineering and TiO2 film modification. J. Mater. Chem. 2011, 21, 6383–6388.

12. Gorer, S.; Hodes, G. Quantum size effects in the study of chemical solution deposition mechanisms of semiconductor films. J. Phys. Chem. 1994, 98, 5338–5346.

13. Moreels, I.; Lambert, K.; de-Muynck, D.; Vanhaecke, F.; Poelman, D.; Martins, J.C.; Allan, G.; Hens, Z. Composition and size-dependent extinction coefficient of colloidal PbSe quantum dots. Chem. Mater. 2007, 19, 6101–6106.

14. Shaller, R.D.; Klimov, V.I. High efficiency carrier multiplication in PbSe nanocrystals: Implications for solar energy conversion. Phys. Rev. Lett. 2004, 92.

15. Michele, P.; Sven, R.; Adam, G.; David, A.K.; Hannah, N.B.; Paolo, M.S.; Daniela, N.; Rodrigo, M.; Assaf, Y.A.; Arie, Z.; et al. TiO2/Cu2O all-oxide heterojunction solar cells produced by spray pyrolysis. Sol. Energy Mater. Sol. Cells 2015, 132, 549–556.

16. Hu, L.H. Microstructure design of nanoporous TiO2 photoelectrodes for dye-sensitized solar cell modules. J. Phys. Chem. B 2007, 111, 358–362.

17. Berger, T.; Lana, V.T.; Monllor, S.D.; Gomez, R. An electrochemical study on the nature of trap states in nanocrystalline rutile thin films. J. Phys. Chem. C 2007, 111, 9936–9942.

18. Qu, J.; Li, G.R.; Gao, X.P. One-dimensional hierarchical titania for fast reaction kinetics of photoanode materials of dye-sensitized solar cells. Energy Environ. Sci. 2010, 3, 2003–2009.

19. Liao, J.Y.; Lei, B.X.; Chen, H.Y.; Kuang, D.B.; Su, C.Y. Oriented hierarchical single crystalline anatase TiO2 nanowire arrays on Ti-foil substrate for efficient flexible dye-sensitized solar cells. Energy Environ. Sci. 2012, 5, 5750–5757.]

20. Liao, J.Y.; Lei, B.X.; Kuang, D.B.; Su, C.Y. Tri-functional hierarchical TiO2 spheres consisting of anatase nanorods and nanoparticles for high efficiency dye-sensitized solar cells. Energy Environ. Sci. 2011, 4, 4079–4085.

21. Ohsaki, Y. Dye-sensitized TiO2 nanotube solar cells: Fabrication and electronic characterization. Phys. Chem. Chem. Phys. 2005, 7, 4157–4163.

22. Zhu, K.; Neale, N.R.; Miedaner, A.; Frank, A.J. Enhanced charge-collection efficiencies and light scattering in dye-sensitized solar cells using oriented TiO2 nanotubes arrays. Nano Lett. 2007, 7, 69–74.

23. Feng, X.; Shankar, K.; Paulose, M.; Grimes, C.A. Tantalum-doped titanium dioxide nanowire arrays for dye-sensitized solar cells with high open-circuit voltage. Angew. Chem. Int. Ed. 2009, 48, 8095–8098.

24. Wu, Q.; Wu, B.X.L.; Hua, S.R.; Cheng, Y.S.; Dai, B.K. Hydrothermal fabrication of hierarchically anatase TiO2 nanowire arrays on FTO glass for dye-sensitized solar cells. Sci. Rep. 2013, 3.

25. Tan, X.; Qiang, P.; Zhang, D.; Cai, X.; Tan, S.; Liu, P.; Mai, W. Three-level hierarchical TiO2nanostructure based high efficiency dye-sensitized solar cells. CrysEngComm 2014, 16, 1020–1025.

26. Lee, Y.L.; Lo, Y.S. Highly efficient quantum-dot-sensitized solar cell based on co-sensitization of CdS/CdSe. Adv. Funct. Mater. 2009, 19, 604–609.

27. Nath, S.S.; Choudhury, M.; Gope, G.; Nath, R.K. PVA embedded ZnO quantum dots for methanol sensing. Nanotrends 2010, 8, 1–4.

28. Kong, E.H.; Chang, Y.J.; Park, Y.C.; Yoon, Y.H.; Park, H.J.; Jang, H.M. Sea urchin TiO2-nanoparticle hybrid composite photoelectrodes for CdS/CdSe/ZnS quantum-dot-sensitized solar cells. Phys. Chem. Chem. Phys. 2012, 14, 4620–4625.

29. Wu, S.J.; Chen, C.Y.; Chen, J.G.; Li, J.Y.; Tung, Y.L.; Ho, K.C.; Wu, C.G. Carbazole containing Ru-based photo-sensitizer for dye-sensitized solar cell. J. Chin. Chem. Soc. 2010, 57, 1127–1130.

30. Rosenblut, M.L.; Lewis, N.S. Ideal behavior of the open circuit voltage of semiconductor liquid junction. J. Phys. Chem. 1989, 93, 3735–3740.

31. Hagfeldt, A.; Lindström, H.; Södergren, S.; Lindquist, S.E. Photoelectrochemical studies of colloidal TiO2 films: The effect of oxygen studied by photocurrent transients. J. Electroanal. Chem. 1995, 381, 39–46.

32. Sheng, X.; Zhao, Y.; Zhai, J.; Jiang, L.; Zhu, D. Electro-hydrodynamic fabrication of ZnO-based dye sensitized solar cells. Appl. Phys. A 2007, 87, 715–719.

33. Sogabe, T.; Shoji, Y.; Ohba, M.; Yoshida, K.; Tamaki, R.; Hong, H.F.; Wu, C.H.; Kuo, C.T.; Tomic, S.; Okada, Y.; et al. Intermediate band dynamic of quantum dots solar cell in concentrated photovoltaic modules. Sci. Rep. 2014, 4.

34. Breitenstein, O.; Langenkamp, M.; Rakotoniaina, J.P.; Zettner, J. The imaging of shunts in solar cells by infrared lock-in thermography. In Proceedings of the 17th European Photovoltaic Solar Energy Conference, Munich, Germany, 22–26 October 2001; pp. 1499–1502.

13 CHAPTER

INVESTIGATION OF THE INAS/GAAS QUANTUM DOTS' SIZE: DEPENDENCE ON THE STRAIN REDUCING LAYER'S POSITION

Manel Souaf [1], Mourad Baira [1], Olfa Nasr [1], Mohamed Helmi Hadj Alouane [1], Hassen Maaref [1], Larbi Sfaxi [1] and Bouraoui Ilahi [2,1,*]

[1] *Laboratoire de Micro-Optoélectronique et Nanostructures, Faculté des Sciences, Avenue de l'environnement, Université de Monastir, Monastir 5019, Tunisia*

[2] *King Saud University, Department of Physics & Astronomy, College of Sciences, P.O. 2455, Riyadh 11451, Saudi Arabia*

ABSTRACT

This work reports on theoretical and experimental investigation of the impact of InAs quantum dots (QDs) position with respect to InGaAs strain reducing layer (SRL). The investigated samples are grown by molecular beam epitaxy and characterized by photoluminescence spectroscopy (PL). The QDs optical transition energies have been calculated by solving the three dimensional Schrödinger equation using the finite element methods and taking into account the strain induced by the lattice mismatch. We have considered a lens shaped InAs QDs in a pure GaAs matrix and either with InGaAs strain reducing cap layer or underlying layer. The correlation between numerical calculation and PL measurements allowed us to track the mean buried QDs size evolution with respect to the surrounding matrix composition. The simulations reveal that the buried QDs' realistic size is less than that experimentally driven from atomic force microscopy observation. Furthermore, the average size is found to be slightly increased for InGaAs capped QDs and dramatically decreased for QDs with InGaAs under layer.

Keywords: modeling; quantum dots; strain reducing layer

1. INTRODUCTION

InAs self-assembled QDs formed by Stranski-Krastanov growth mode have been a subject of intense research activity towards the development of innovative devices such as QDs lasers [1,2], light emitters and detectors [3,4,5,6]. More recently, this kind of QDs has been found to be promising for photovoltaic applications [7]. However, the efficient exploitation of the self-assembled QDs towards high performance devices relay on the best optimization of their size dispersion and control of their emission energy. Accordingly, it is important to understand the impact of the capping and/or the underling layer matrix composition on the evolution of the QDs size and electronic structure. Several methods, based on the strain engineering, have been reported [8,9,10]. Indeed, replacing the GaAs surrounding material by InGaAs alloy leads to partial strain relief-induced modification of confinement potential allowing to manipulate inter and intraband transition energies in InAs QDs [11,12]. The SRL has been employed as a QDs capping layer [13], underling layer [14], or a combination between these two approaches [15]. Despite the numerous theoretical and experimental reports on the impact of the SRL on the optical and structural properties of the QDs [16,17,18,19,20,21,22,23], less attention has been devoted to the combined theoretical and optical spectroscopy investigation of the impact of the InAs QDs position with respect to the SRL on the evolution of the buried dots' size. Actually, it is well established that the buried InAs QDs undergo a compositional change duo to In segregation and in-situ intermixing [20,24,25]. Accordingly, the accurate modeling of the QDs electronic properties requires the knowledge of the exact In compositional profile [23,26]. In the meanwhile, with the absence of such data, the combination of photoluminescence spectroscopy and numerical calculation of the InAs QDs' electronic structure could constitute an alternative solution allowing a comprehensive investigation of the asymptotic buried dots' size evolution depending on the surrounding materials composition.

In this context, we report on a numerical and experimental investigation of the effect of strain reducing layer position on the size and electronic properties of the InAs/GaAs QDs system. The theoretical estimation explains well the **observed experimental results.**

2. RESULTS AND DISCUSSION

Solving the Schrödinger equation with single band effective mass approximation by using the finite element methods (FEM) provided by COMSOL Multiphysics software, [27] affords a numerical estimation of the energy levels and corresponding wave functions. In this calculation, the InAs QDs are considered to be lens shaped as shown by the Figure 1.

The Schrödinger equation in cylindrical coordinates is given by:

$$\frac{-\hbar^2}{2m^*}\left(\frac{\partial^2}{\partial r^2}+\frac{1}{r}\frac{\partial}{\partial r}+\frac{1}{r^2}\frac{\partial^2}{\partial \theta^2}+\frac{\partial^2}{\partial z^2}\right)\psi(r,\theta,z)+V\psi(r,\theta,z)=E\psi(r,\theta,z) \tag{1}$$

where E represents the electron (hole) energy of state, m* electron (hole) effective masse, ħ the reduced Planck constant, θ is the azimuthal angle varied from 0 to 2π, r and z are, respectively, the radial and axial coordinates. Due to the rotational symmetric of the system, the Schrödinger equation can be reduced to a two dimensional problem were ψ(r,θ,z) can be written in the following form:

$$\psi(\mathrm{r}, \phi, \mathrm{z}) = \chi(\mathrm{r}, \mathrm{z})\Phi(\theta) \tag{2}$$

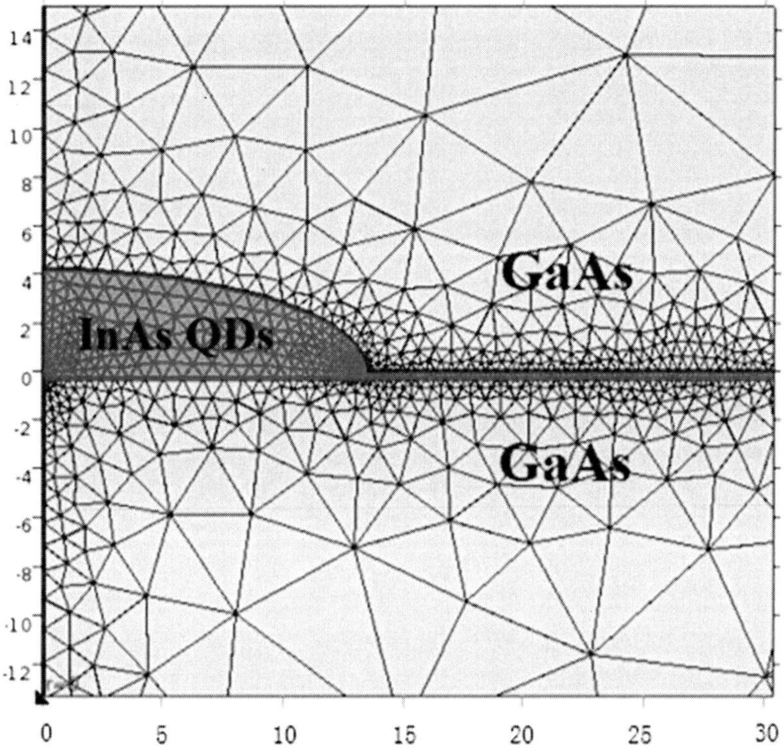

Figure 1. Schematic presentation of the zoomed lens shaped InAs quantum dots (QD) in a GaAs matrix. Also shown is the adaptive mesh refinement around the QD and the wetting layer.

By dividing Equations (1) by (2) and then multiplying by r2, we obtain:

$$\frac{-\hbar^2}{2m^*}\left(\frac{r^2}{\chi(\mathrm{r},\mathrm{z})}\frac{\partial^2\chi(\mathrm{r},\mathrm{z})}{\partial r^2} + \frac{r}{\chi(\mathrm{r},\mathrm{z})}\frac{\partial\chi(\mathrm{r},\mathrm{z})}{\partial r} + + \frac{\partial^2\chi(\mathrm{r},\mathrm{z})}{\partial z^2}\right) + (V-E)r^2 = \frac{\hbar^2}{2m^*}\frac{1}{\Phi(\theta)}\frac{\partial^2\Phi(\theta)}{\partial\theta^2} \tag{3}$$

the right part can be written as follows:

$$\frac{\hbar^2}{2m^*}\frac{1}{\Phi(\theta)}\frac{\partial^2\Phi(\theta)}{\partial\theta^2}=\frac{-\hbar^2}{2m^*}n^2 \tag{4}$$

where n is the orbital quantum number that must be integer:

$$\frac{-\hbar^2}{2m^*}\left(\frac{r^2}{\chi(r,z)}\frac{\partial^2\chi(r,z)}{\partial r^2}+\frac{r}{\chi(r,z)}\frac{\partial\chi(r,z)}{\partial r}++\frac{\partial^2\chi(r,z)}{\partial z^2}\right)+(V-E)r^2+\frac{\hbar^2}{2m^*}n^2=0 \tag{5}$$

when multiplying this equation by χ(r,z)/r2, we find the following equation:

$$-\frac{\hbar^2}{2m^*}\left(\frac{\partial^2\chi(r,z)}{\partial r^2}+\frac{\partial^2\chi(r,z)}{\partial z^2}\right)+\left(V+\frac{1}{r^2}\frac{\hbar^2}{2m^*}n^2\right)\chi(r,z)-\frac{\hbar^2}{2m^*}\frac{1}{r}\frac{\partial\chi(r,z)}{\partial r}=E\chi(r,z) \tag{6}$$

To calculate the energy states and theirs wave functions for electron/hole in the QD, Dirichlet conditions are applied where the wave function must be zero (outside the considered domain) and the Neumann conditions are employed for the internal boundaries to ensure the continuity of the wave function. One of the major advantages of FEM is the flexibility to vary the desired precision over the entire domain by controlling the quality of the mesh. Accordingly, an adaptive mesh refinement has been employed as shown by the Figure 1.

The impact of strain on the carrier confinement due to the lattice mismatch between InAs and GaAs materials is also taken into account for the calculation [20]. Hydrostatic and uniaxial strains are given by the following expressions:

$$\epsilon_h=\epsilon_{xx}+\epsilon_{yy}+\epsilon_{zz} \tag{7}$$

and

$$\epsilon_b=\epsilon_{zz}-\frac{1}{2}(\epsilon_{xx}+\epsilon_{yy}) \tag{8}$$

where εb, εh are hydrostatics and uniaxial strain respectively. The lattice mismatch between InAs and GaAs is given by:

$$\epsilon_{xx} = \epsilon_{yy} = \frac{a_{GaAs} - a_{InAs}}{a_{InAs}} \tag{9}$$

and

$$\epsilon_{zz} = -2\frac{C_{12}}{C_{11}}\epsilon_{xx} \tag{10}$$

The strained effect on conduction band is $\delta E_c = a_c \epsilon_h$ and for valence band is

$$\delta E_v = a_v \epsilon_h - \frac{1}{2} b \epsilon_b \tag{11}$$

then the strained band gap energy (Eg) can be written as:

$$\mathrm{Eg}_{\mathrm{InAs-str}} = \mathrm{Eg}_{\mathrm{InAs-unstr}} + \delta E_c - \delta E_v \tag{12}$$

where EgInAs-unstr is the unstrained InAs band gap energy and EgInAs-str presents the strained one. The parameters used in this numerical approach are summarized in Table 1 [28].

Table 1. InAs and GaAs parameters used for the calculation.

Material	m_e^* (m_0)	m_h^* (m_0)	C_{11} 10^{11} (dyn cm^{-2})	C_{12} 10^{11} (dyn cm^{-2})	($a_c - a_v$) (eV)	b (eV)
InAs	0.023	0.41	8.32	4.52	−4.08	−1.8
GaAs	0.068	0.5	12.21	5.66	−6	−2

The single particle optical transition energies Ei can be deduced by the following equation:

$$E_i = e_i + Eg_{InAs-Str} + h_i \tag{13}$$

where ei (hi) correspond to the electron (hole) confined state.

Figure 2 shows the high excitation power 12 K PL data recorded for the reference sample where the QDs are in a pure GaAs matrix. The appearance of multiple peaks results from the states filling effects. The ground state (GS) emission energy is located around 1.086 eV with two excited states at 1.155 eV and 1.224 eV. The appearance of well resolved optical transition energies allows us to estimate the mean size of the buried QDs. Indeed, by tuning the QDs AR, it is possible to fit the calculated ground and excited states transition energies to the experimentally measured PL emission energies.

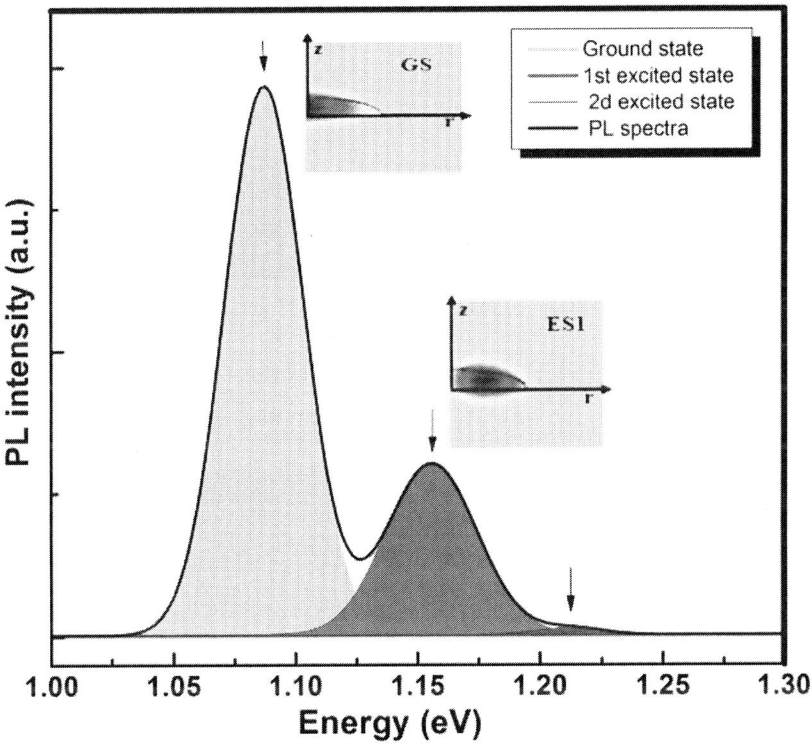

Figure 2. 12 K photoluminescence spectroscopy (PL) spectra taken with an excitation power of 400 mW, from the reference InAs QDs in a pure GaAs matrix showing a ground state and two excited states emission peaks. The arrows indicate the calculated transition energies. The electron envelope functions are also shown as insets.

The calculation has been based on the three emission energies appearing in the PL spectrum. The experimental and calculated emission energies are perfectly matched within some meV yielding an average QD height of 3.8 nm and a base diameter around 28 nm. The calculated wave functions of the first two confined electron states using the estimated realistic size for the buried QDs with an InAs wetting layer thickness of 0.3 nm are shown in the Figure 2. It is important to notice that atomic force microscopy (AFM) investigation of similar uncapped structure revealed a surface density around 2.4 × 1010 cm−2 and a mean dots' height of 8 nm with a base diameter around 35 nm giving rise to an aspect ratio (AR) of 0.23. This shows that the values are over estimated. Indeed, the QDs height can be reduced by the capping process [29,30]. Additionally, the AFM technique may also provide an over estimated size due to the tip artifact [31,32]. It is also reasonable to take into account the fact that assuming a pure InAs QDs in the calculation might yields smaller size QDs than the real case where small amount of In atoms may out diffuse.

In order to study the impact the QDs position with respect to the InGaAs SRL, we have conducted a first investigation using these reference QDs parameters. This implies that the InGaAs capping or underlying layer will contribute to the strain reduction without affecting the QDs size. The calculations expect a ground state emission energy redshift around 20 meV upon capping the InAs QDs by 10 nm of InGaAs and around 10 meV when the SRL is inserted below the QDs.

The PL spectra show that capping the QDs by 10 nm (QDCL) leads to a decrease of the PL emission energy by approximately 32 meV as shown by the Figure 3.

The excess energy red shift could arise from the increase of the QDs' size a result of the strain driven phase separation of the capping layer alloy. This result is in agreement with recently reported structural investigation on similar structures [23]. In the meanwhile, the GS emission energy of the InAs QDs with InGaAs underlying layer (QDUL) shows an unexpected blueshift of the GS emission energy up to 1.204 eV with broadening of the PL linewidth and reduction of the intersublevel spacing energy. A shown by the inset of the Figure 3,

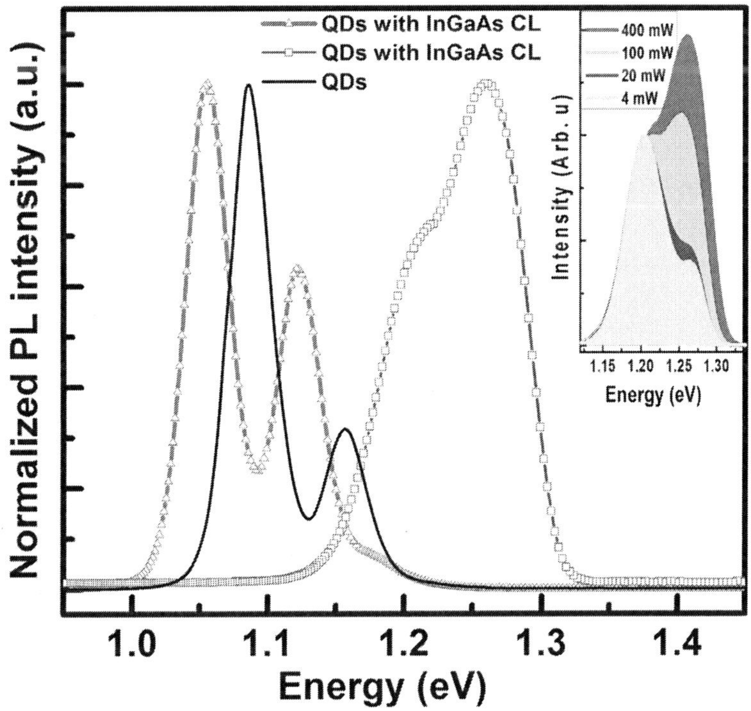

Figure 3. 12 K PL spectra taken with an excitation power of 400 mW, from the reference InAs QDs in a pure GaAs matrix and from the InAs QDs with SRL as a cap layer (empty triangles) and as an underlying layer (empty squares). The Inset shows the excitation power dependent PL spectra from the InAs QDs with InGaAs underlying layer. Multiple peaks in each spectra arise from the stat filling effects.

the persistence of the excited states emission energy peak at the lower excitation power, testify to a reduced number of coherent QDs [11]. Furthermore, AFM investigation on similar uncapped InAs QDs revealed an increased dots surface density to 4 × 1010 cm−2, with a broadened size distribution a reduced mean dots' height down to 5 nm without appearance of bimodal size distribution [11] confirming the later statement. These observations indicate that the mean dots' size has been seriously altered by the InGaAs underlying layer.

In fact, by varying the dots' size we succeed in accurately fitting the PL emission energies by the calculated values for both InAs QDs with either SRL capping or underlying layer. The results are summarized in Table 2, where the estimated results using the reference QDs size are shown together with the simulated real dots' size obtained by tuning the dots' size to match the experimental transition energies. Accordingly, the calculation reveals an increased buried dots' size upon capping by InGaAs SRL, while keeping the same aspect ratio. This change of InAs QDs size is explained by the redistribution of Indium and Gallium atoms during the deposition of the InGaAs capping layer on InAs QDs resulted from alloy phase separation phenomena [16].

Table 2. Experimental and theoretical Ground state and first two excited states optical transition energies as well as corresponding QD size. The experimental emission energies are given by PL spectroscopy and the average QDs size is estimated through AFM. The expected results refer to the calculated transition energies keeping the simulated buried dot size for the reference sample. The Simulation refers to the realistic dots' size driven by fitting the theoretical emission energies to the corresponding PL peaks.

Sample	Data	H_{QD}/D_{QD}	Aspect Ratio (α)	E_0 (eV)	E_1 (eV)	E_2 (eV)
QDs	Experiments	8.5/35	0.24	1.086	1.156	1.224
	Simulation	3.8/28	0.13	1.083	1.151	1.238
QDCL	Experiments	8.5/35	0.24	1.054	1.117	1.194
	Simulation	4/30	0.13	1.051	1.113	1.193
	Expected results	3.8/28	0.13	1.068	1.131	1.213
QDUL	Experiments	5/41	0.12	1.21	1.261	–
	Simulation	2.5/29	0.08	1.204	1.265	1.331
	Expected results	3.8/28	0.13	1.071	1.141	1.217

However, a drastic change of the mean size and aspect ratio is found to occur for the QDs with InGaAs underlying layer. Indeed, the dots' height is found to decrease to 2.5 nm and the base diameter is increased to 29 nm. In fact, indium segregation during the growth of the SRUL is expected to increase the nominal thickness of InAs material from 2.4 ML to up to 3 ML. Indeed, when the InAs nominal thickness exceeds a critical value, coalesced or/and dislocated QDs are formed and the number and size of active QDs could be critically reduced. This effect is confirmed by detailed PL investigation indicating a degradation of the QDs structural properties [11]. Additionally, the reduced strain increases the InAs critical thickness which is favorable to the formation of thicker wetting layers allowing the growth of smaller size QDs with higher areal density [33].

To further investigate the effect of the size modification that occurred following the growth of the QDs on the top of the InGaAs layer, the calculated electrons wave functions for the ground state and the first excited state are shown by the Figure 4 in the (r,z) plan.

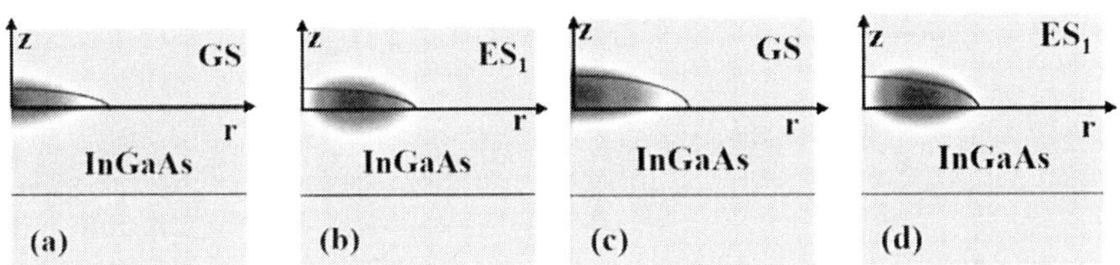

Figure 4. (a) and (b) indicates the envelop function for electrons with the simulated real size of the buried dots; (c) and (d) refer to the envelop functions expected for QDs with InGaAs underlying layer (QDUL) with a dot-size that is supposed to be the same as the reference sample.

Indeed, as a consequence of the decreased realistic size, compared to that of the reference QDs, the electron wave function is found to spread out of the QDs. This behavior indicates less localization as a consequence of the lowering of the effective confining potential barrier and increased confined energy levels.

3. EXPERIMENTAL SECTION

For experimental validation and deeper investigation of the effects of the InGaAs layer's position on the structural and optical properties of InAs QDs we have used a single InAs QDs layer as reference sample. In the two other samples, the strain has been reduced by an InGaAs layer with 15% indium composition. In the QDCL sample, the QDs were covered by 10 nm InGaAs. However, in the QDUL sample, the InAs QDs layer was deposited on a 10 nm InGaAs. The samples are characterized by PL spectroscopy using a conventional lock-in technique. A 514.5 nm line of an Ar + laser with 0.2 mm spot diameter has been employed as excitation source. More details on the growth process, optical and morphological properties of these samples can be found elsewhere [11].

4. CONCLUSIONS

In summary, a combination of a simple numerical approach in the frame of the effective mass approximation method and photoluminescence spectroscopy have been employed to perform a detailed investigation of the effect of the InAs QDs position with respect to the strain reducing layer on the buried dots' size. The numerical calculations using the same dot-size expects a redshift of the emission energies upon underlaying or capping the dots by InGaAs alloy. The PL results show additional redshift for the emission energy of the InGaAs capped QDs and a realistic QDs size simulation revealed an enhancement induced by the alloy phase separation. When the QDs are grown on the top of the InGaAs SRL, the experiments show an unexpected blueshift. The simulation of the buried dots' size indicated a drastic reduction of the QDs' height leading to less carrier localization. Our results provide insight towards better understanding and optimization of QDs' properties by using SRL.

ACKNOWLEDGMENTS

The last author (Bouroui Ilahi) would like to extend his sincere appreciation to the Deanship of Scientific Research at king Saud University for funding this Research group NO: RG-1436-014.

AUTHOR CONTRIBUTIONS

Manel Souaf and Mourad Baira did the calculations, Olfa Nasr and Mohamed Helmi Hadj Alouane has performed the Photoluminescence measurements, Larbi Sfaxi and Hassen Maaref did the samples growth. Bouraoui Ilahi has planned the work and contributed to the characterization and numerical calculation. All the co-authors have contributed to the revision of the paper written by Manel Souaf.

REFERENCES

1. Majid, M.A.; Childs, D.T.D.; Kennedy, K.; Airey, R.; Hogg, R.A.; Clarke, E.; Spencer, P.; Murray, R. O-band excited state quantum dot bilayer lasers. App. Phys. Lett. 2011, 99, 051101.]
2. Ustinov, V.M.; Zhukov, A.E. GaAs-based long-wavelength lasers. Semicond. Sci. Technol. 2000, 15, 41–54.
3. Salter, C.L.; Stevenson, R.M.; Farrer, I.; Nicoll, C.A.; Ritchie, D.A.; Shields, A.J. Single-photon and entangled-photon sources for quantum information. Nature 2010, 465, 594–597.
4. Shields, A.J.; O'Sullivan, M.P.; Farrer, I.; Ritchie, D.A.; Hogg, R.A.; Leadbeater, M.L.; Norman, C.E.; Pepper, M. Detection of single photons using a field-effect transistor gated by a layer of quantum dots. Appl. Phys. Lett. 2000, 76, 3673–3675.
5. Ledentsov, N.N. Quantum dot laser. Semicond. Sci. Technol. 2011, 26, 014001.
6. Chen, J.X.; Markus, A.; Fiore, A.; Oesterle, U.; Stanley, R.P.; Carlin, J.F.; Houdré, R.; Ilegems, M.; Lazzarini, L.; Nasi, L.; et al. Tuning InAs/GaAs quantum dot properties under Stranski-Krastanov growth mode for 1.3 μm applications. J. Appl. Phys. 2002, 91, 6710–6716.

7. Sogabe, T.; Shoji, Y.; Ohba, M.; Yoshida, K.; Tamaki, R.; Hong, H.F.; Wu, C.H.; Kuo, C.T.; Tomić, S.; Okada, Y. Intermediate-band dynamics of quantum dots solar cell in concentrator photovoltaic modules. Sci. Rep. 2014, 4.

8. Kong, L.; Sun, W.; Feng, Z.C.; Xie, S.; Zhou, Y.; Wang, R.; Zhang, C.; Zong, Z.; Wang, H.; Qiao, Q.; et al. Effects of InAlAs strain reducing layer on the photoluminescence properties of InAs quantum dots embedded in InGaAs/GaAs quantum wells. Thin Solid Films 2014, 562, 440–444.

9. Liu, H.Y.; Hopkinson, M.; Harrison, C.N.; Steer, M.J.; Frith, R.; Sellers, I.R.; Mowbray, D.J.; Skolnick, M.S. Optimizing the growth of 1.3 μm InAs/InGaAs dots-in-a-well structure. J. Appl. Phys. 2003, 93, 2931–2936.

10. Seravalli, L.; Minelli, M.; Frigeri, P.; Franchi, S.; Guizzetti, G.; Patrini, M.; Ciabattoni, T.; Geddo, M. Quantum dot strain engineering of InAs/InGaAs nanostructures. J. Appl. Phys. 2007, 101.

11. Nasr, O.; HadjAlouane, M.H.; Maaref, H.; Hassen, F.; Sfaxi, L.; Ilahi, B. Comprehensive investigation of optical and electronic properties of tunable InAs QDs optically active at O-band telecommunication window with (In)GaAs surrounding material. J. Lumin. 2014, 148, 243–248.

12. Ustinov, V.M.; Maleev, N.A.; Zhukov, A.E.; Kovsh, A.R.; Egorov, A.Yu.; Lunev, A.V.; Volovik, B.V.; Krestnikov, I.L.; Musikhin, Yu.G.; Bert, N.A.; et al. InAs/InGaAs quantum dot structures on GaAs substrates emitting at 1.3 μm. Appl. Phys. Lett. 1999, 74, 1384–1385.

13. Nishi, K.; Saito, H.; Sugou, S.; Lee, J.S. A narrow photoluminescence linewidth of 21 meV at 1.35 μm from strain-reduced InAs quantum dots covered by In0.2Ga0.8 as grown on GaAs substrates. Appl. Phys. Lett. 1999, 74, 1111–1113.

14. Pires, M.P.; Landi, S.M.; Tribuzy, C.V.B.; Nunes, L.A.; Marega, E.; Souza, P.L. InAs quantum dots over InGaAs for infrared photodetectors. J. Cryst. Growth 2004, 272, 192–197.

15. Chen, R.; Liu, H.Y.; Sun, H.D. Electronic energy levels and carrier dynamics in InAs/InGaAs dots-in-a-well structure investigated by optical spectroscopy. J. Appl. Phys. 2010, 107.

16. Joyce, P.B.; Kryzewski, T.J.; Bell, G.R.; Joens, T.S.; le Ru, E.C.; Murray, R. Optimizing the growth of 1.3 μm InAs/GaAs quantum dots. Phys. Rev. B 2001, 64, 235317.

17. Le Ru, E.C.; Howe, P.; Joens, T.S.; Murray, R. Strain-engineered InAs/GaAs quantum dots for long-wavelength emission. Phys. Rev. B 2003, 67, 165303.

18. Nabetani, Y.; Matsumoto, T.; Sasikala, G.; Suemune, I. Theory of strain states in InAs quantum dots and dependence on their capping layers. J. Appl. Phys. 2005, 98, 063502.

19. Ilahi, B.; Sfaxi, L.; Hassen, F.; Maaref, H.; Salem, B.; Guillot, G.; Jbeli, A.; Marie, X. Optical properties of 1.3 μm room temperature emitting InAs quantum dots covered by In0.4Ga0.6 As/GaAs hetero-capping layer. Appl. Phys. A 2005, 81, 813–816.

20. Cherkashin, N.; Reboh, S.; Htch, M.J.; Claverie, A.; Preobrazhenskii, V.V.; Putyato, M.A.; Semyagin, B.R.; Chaldyshev, V.V. Determination of stress, strain, and elemental distribution within In(Ga)As quantum dots embedded in GaAs using advanced transmission electron microscopy. Appl. Phys. Lett. 2013, 102, 173115.

21. Ilahi, B.; Sfaxi, L.; Maaref, H. Optical investigation of InGaAs-capped InAs quantum dots: Impact of the strain-driven phase separation and dependence upon post-growth thermal treatment. J. Lumin. 2007, 127, 741–746.

22. Maia, A.D.B.; da Silva, E.C.F.; Quivy, A.A.; Bindilatti, V.; de Aquino, V.M.; Dias, I.F.L. Simulation of the electronic properties of InxGa1−x As quantum dots and their wetting layer under the influence of indium segregation. J. Appl. Phys. 2013, 114, 083708.

23. Wang, P.; Bleloch, A.L.; Falke, M.; Goodhew, P.J.; Ng, J.; Missous, M. Direct measurement of composition of buried quantum dots using aberration-corrected scanning

transmission electron microscopy. Appl. Phys. Lett. 2006, 89, 072111.

24. García, J.M.; Medeiros-Ribeiro, G.; Schmidt, K.; Ngo, T.; Feng, J.L.; Lorke, A.; Kotthaus, J.; Petroff, P.M. Intermixing and shape changes during the formation of InAs self-assembled quantum dots. Appl. Phys. Lett. 1997, 74, 2014–2016.

25. Joyce, P.B.; Krzyzewski, T.J.; Bell, G.R.; Jones, T.S. Surface morphology evolution during the overgrowth of large InAs-GaAs quantum dots. Appl. Phys. Lett. 2001, 79, 3615–3617.

26. Kegel, I.; Metzger, T.H.; Lorke, A.; Peisl, J.; Stangl, J.; Bauer, G.; Nordlund, K.; Schoenfeld, W.V.; Petroff, P.M. Determination of strain fields and composition of self-organized quantum dots using x-ray diffraction. Phys. Rev. B 2001, 63, 035318.

27. COMSOL Multiphysics. Available online: http://www.comsol.com (accessed on 17 July 2015).

28. Vurgaftman, I.; Meyer, J.R.; Ram-Mohan, L.R. Band parameters for III–V compound semiconductors and their alloys. J. Appl. Phys. 2001, 89, 5815–5875.

29. Grossi, D.F.; Smereka, P.; Keizer, J.G.; Ulloa, J.M.; Koenraad, P.M. Height control of self-assembled quantum dots by strain engineering during capping. Appl Phys. Lett. 2014, 105, 143104.

30. Kim, J.S.; Yu, P.W.; Lee, J.S.; Kim, J.S.; Kim, S.G.; Leem, J.; Jeon, M. Height-controlled InAs quantum dots by using a thin InGaAs layer. Appl. Phys. Lett. 2002, 80, 4714–4716.

31. Shiramine, K.; Muto, S.; Shibayama, T.; Sakaguchi, N.; Ichinose, H.; Kozaki, T.; Sato, S.; Nakata, Y.; Yokoyama, N.; Taniwaki, M. Tip artifact in atomic force microscopy observations of InAs quantum dots grown in Stranski-Krastanow mode. J. Appl. Phys. 2007, 101, 033527.

32. Gong, Q.; Liang, J.B.; Xu, B.; Ding, D.; Li, H.X.; Jiang, C.; Zhou, W.; Liu, F.Q.; Wang, Z.G.; Qiu, X.H.; et al. Analysis of atomic force microscopic results of InAs islands formed by molecular beam epitaxy. J. Cryst. Growth 1998, 192, 376–380.

33. Simmonds, P.J.; Sun, M.; Laghumavarapu, R.B.; Liang, B.; Norman, A.G.; Luo, J.W.; Huffaker, D.L. Improved quantum dot stacking for intermediate band solar cells using strain compensation. Nanotechnology 2014, 25, 445402.

INDEX

B

bioanalysis, 136, 147

D

decomposition, 21, 247, 248, 249, 250, 251, 254, 257, 258

F

fluorescence, 48, 53, 124, 125, 228, 235, 237, 241, 245

G

GaN quantum, 247, 248, 256

H

hierarchical, 259, 260, 261, 274

I

imaging, 36, 45, 123, 124, 126, 241, 242, 243, 244, 245, 275

L

Low density, 248

M

modeling, 278, 279

N

Nitrogen-doped, 48, 61

P

photodynamic, 59, 135, 136, 244, 245

photoluminescence, 35, 36, 288, 290
probe, 76, 132, 142, 147, 222

Q

Quantum dot solar, 48
Quasi-blue emission, 36

S

Self-assembled, 156, 203, 204, 216
semiconductor, 1, 2, 3, 4, 7, 8, 256, 273, 275
Silicon quantum dots, 36
simulation, 64, 66, 73, 288
spectroscopy, 40, 50, 53, 54, 283, 286, 288, 291
strain reducing layer, 277, 278, 279, 288, 290

T

toxicology, 136

W

Water-soluble, 36, 44, 45, 46, 60